Aberdeenshire
COUNCIL
Aberdeenshire Libraries
www.aberdeenshire.gov.uk/libraries
25 APR 2017
WITHDRAWN
FROM LIBRARY
ABERDEENSHIRE
LIBRARIES

D1421484
3202406

VERY SHORT INTRODUCTIONS are for anyone wanting a stimulating and accessible way into a new subject. They are written by experts, and have been translated into more than 45 different languages.

The series began in 1995, and now covers a wide variety of topics in every discipline. The VSI library now contains over 500 volumes—a Very Short Introduction to everything from Psychology and Philosophy of Science to American History and Relativity—and continues to grow in every subject area.

Very Short Introductions available now:

ACCOUNTING Christopher Nobes
ADOLESCENCE Peter K. Smith
ADVERTISING Winston Fletcher
AFRICAN AMERICAN RELIGION Eddie S. Glaude Jr
AFRICAN HISTORY John Parker and Richard Rathbone
AFRICAN RELIGIONS Jacob K. Olupona
AGEING Nancy A. Pachana
AGNOSTICISM Robin Le Poidevin
AGRICULTURE Paul Brassley and Richard Soffe
ALEXANDER THE GREAT Hugh Bowden
ALGEBRA Peter M. Higgins
AMERICAN HISTORY Paul S. Boyer
AMERICAN IMMIGRATION David A. Gerber
AMERICAN LEGAL HISTORY G. Edward White
AMERICAN POLITICAL HISTORY Donald Critchlow
AMERICAN POLITICAL PARTIES AND ELECTIONS L. Sandy Maisel
AMERICAN POLITICS Richard M. Valelly
THE AMERICAN PRESIDENCY Charles O. Jones
THE AMERICAN REVOLUTION Robert J. Allison
AMERICAN SLAVERY Heather Andrea Williams
THE AMERICAN WEST Stephen Aron
AMERICAN WOMEN'S HISTORY Susan Ware
ANAESTHESIA Aidan O'Donnell
ANARCHISM Colin Ward
ANCIENT ASSYRIA Karen Radner
ANCIENT EGYPT Ian Shaw
ANCIENT EGYPTIAN ART AND ARCHITECTURE Christina Riggs
ANCIENT GREECE Paul Cartledge
THE ANCIENT NEAR EAST Amanda H. Podany
ANCIENT PHILOSOPHY Julia Annas
ANCIENT WARFARE Harry Sidebottom
ANGELS David Albert Jones
ANGLICANISM Mark Chapman
THE ANGLO-SAXON AGE John Blair
ANIMAL BEHAVIOUR Tristram D. Wyatt
THE ANIMAL KINGDOM Peter Holland
ANIMAL RIGHTS David DeGrazia
THE ANTARCTIC Klaus Dodds
ANTISEMITISM Steven Beller
ANXIETY Daniel Freeman and Jason Freeman
THE APOCRYPHAL GOSPELS Paul Foster
ARCHAEOLOGY Paul Bahn
ARCHITECTURE Andrew Ballantyne
ARISTOCRACY William Doyle
ARISTOTLE Jonathan Barnes
ART HISTORY Dana Arnold
ART THEORY Cynthia Freeland
ASIAN AMERICAN HISTORY Madeline Y. Hsu
ASTROBIOLOGY David C. Catling
ASTROPHYSICS James Binney

ATHEISM Julian Baggini
AUGUSTINE Henry Chadwick
AUSTRALIA Kenneth Morgan
AUTISM Uta Frith
THE AVANT GARDE David Cottington
THE AZTECS Davíd Carrasco
BABYLONIA Trevor Bryce
BACTERIA Sebastian G. B. Amyes
BANKING John Goddard and John O. S. Wilson
BARTHES Jonathan Culler
THE BEATS David Sterritt
BEAUTY Roger Scruton
BEHAVIOURAL ECONOMICS Michelle Baddeley
BESTSELLERS John Sutherland
THE BIBLE John Riches
BIBLICAL ARCHAEOLOGY Eric H. Cline
BIOGRAPHY Hermione Lee
BLACK HOLES Katherine Blundell
BLOOD Chris Cooper
THE BLUES Elijah Wald
THE BODY Chris Shilling
THE BOOK OF MORMON Terryl Givens
BORDERS Alexander C. Diener and Joshua Hagen
THE BRAIN Michael O'Shea
THE BRICS Andrew F. Cooper
THE BRITISH CONSTITUTION Martin Loughlin
THE BRITISH EMPIRE Ashley Jackson
BRITISH POLITICS Anthony Wright
BUDDHA Michael Carrithers
BUDDHISM Damien Keown
BUDDHIST ETHICS Damien Keown
BYZANTIUM Peter Sarris
CALVINISM Jon Balserak
CANCER Nicholas James
CAPITALISM James Fulcher
CATHOLICISM Gerald O'Collins
CAUSATION Stephen Mumford and Rani Lill Anjum
THE CELL Terence Allen and Graham Cowling
THE CELTS Barry Cunliffe
CHAOS Leonard Smith
CHEMISTRY Peter Atkins
CHILD PSYCHOLOGY Usha Goswami
CHILDREN'S LITERATURE Kimberley Reynolds
CHINESE LITERATURE Sabina Knight
CHOICE THEORY Michael Allingham
CHRISTIAN ART Beth Williamson
CHRISTIAN ETHICS D. Stephen Long
CHRISTIANITY Linda Woodhead
CITIZENSHIP Richard Bellamy
CIVIL ENGINEERING David Muir Wood
CLASSICAL LITERATURE William Allan
CLASSICAL MYTHOLOGY Helen Morales
CLASSICS Mary Beard and John Henderson
CLAUSEWITZ Michael Howard
CLIMATE Mark Maslin
CLIMATE CHANGE Mark Maslin
COGNITIVE NEUROSCIENCE Richard Passingham
THE COLD WAR Robert McMahon
COLONIAL AMERICA Alan Taylor
COLONIAL LATIN AMERICAN LITERATURE Rolena Adorno
COMBINATORICS Robin Wilson
COMEDY Matthew Bevis
COMMUNISM Leslie Holmes
COMPLEXITY John H. Holland
THE COMPUTER Darrel Ince
COMPUTER SCIENCE Subrata Dasgupta
CONFUCIANISM Daniel K. Gardner
THE CONQUISTADORS Matthew Restall and Felipe Fernández-Armesto
CONSCIENCE Paul Strohm
CONSCIOUSNESS Susan Blackmore
CONTEMPORARY ART Julian Stallabrass
CONTEMPORARY FICTION Robert Eaglestone
CONTINENTAL PHILOSOPHY Simon Critchley
COPERNICUS Owen Gingerich
CORAL REEFS Charles Sheppard
CORPORATE SOCIAL RESPONSIBILITY Jeremy Moon
CORRUPTION Leslie Holmes
COSMOLOGY Peter Coles
CRIME FICTION Richard Bradford
CRIMINAL JUSTICE Julian V. Roberts
CRITICAL THEORY Stephen Eric Bronner
THE CRUSADES Christopher Tyerman

CRYPTOGRAPHY Fred Piper and Sean Murphy
CRYSTALLOGRAPHY A. M. Glazer
THE CULTURAL REVOLUTION Richard Curt Kraus
DADA AND SURREALISM David Hopkins
DANTE Peter Hainsworth and David Robey
DARWIN Jonathan Howard
THE DEAD SEA SCROLLS Timothy Lim
DECOLONIZATION Dane Kennedy
DEMOCRACY Bernard Crick
DEPRESSION Jan Scott and Mary Jane Tacchi
DERRIDA Simon Glendinning
DESCARTES Tom Sorell
DESERTS Nick Middleton
DESIGN John Heskett
DEVELOPMENTAL BIOLOGY Lewis Wolpert
THE DEVIL Darren Oldridge
DIASPORA Kevin Kenny
DICTIONARIES Lynda Mugglestone
DINOSAURS David Norman
DIPLOMACY Joseph M. Siracusa
DOCUMENTARY FILM Patricia Aufderheide
DREAMING J. Allan Hobson
DRUGS Les Iversen
DRUIDS Barry Cunliffe
EARLY MUSIC Thomas Forrest Kelly
THE EARTH Martin Redfern
EARTH SYSTEM SCIENCE Tim Lenton
ECONOMICS Partha Dasgupta
EDUCATION Gary Thomas
EGYPTIAN MYTH Geraldine Pinch
EIGHTEENTH-CENTURY BRITAIN Paul Langford
THE ELEMENTS Philip Ball
EMOTION Dylan Evans
EMPIRE Stephen Howe
ENGELS Terrell Carver
ENGINEERING David Blockley
ENGLISH LITERATURE Jonathan Bate
THE ENLIGHTENMENT John Robertson
ENTREPRENEURSHIP Paul Westhead and Mike Wright
ENVIRONMENTAL ECONOMICS Stephen Smith
ENVIRONMENTAL POLITICS Andrew Dobson
EPICUREANISM Catherine Wilson
EPIDEMIOLOGY Rodolfo Saracci
ETHICS Simon Blackburn
ETHNOMUSICOLOGY Timothy Rice
THE ETRUSCANS Christopher Smith
EUGENICS Philippa Levine
THE EUROPEAN UNION John Pinder and Simon Usherwood
EVOLUTION Brian and Deborah Charlesworth
EXISTENTIALISM Thomas Flynn
EXPLORATION Stewart A. Weaver
THE EYE Michael Land
FAMILY LAW Jonathan Herring
FASCISM Kevin Passmore
FASHION Rebecca Arnold
FEMINISM Margaret Walters
FILM Michael Wood
FILM MUSIC Kathryn Kalinak
THE FIRST WORLD WAR Michael Howard
FOLK MUSIC Mark Slobin
FOOD John Krebs
FORENSIC PSYCHOLOGY David Canter
FORENSIC SCIENCE Jim Fraser
FORESTS Jaboury Ghazoul
FOSSILS Keith Thomson
FOUCAULT Gary Gutting
THE FOUNDING FATHERS R. B. Bernstein
FRACTALS Kenneth Falconer
FREE SPEECH Nigel Warburton
FREE WILL Thomas Pink
FRENCH LITERATURE John D. Lyons
THE FRENCH REVOLUTION William Doyle
FREUD Anthony Storr
FUNDAMENTALISM Malise Ruthven
FUNGI Nicholas P. Money
GALAXIES John Gribbin
GALILEO Stillman Drake
GAME THEORY Ken Binmore
GANDHI Bhikhu Parekh
GENES Jonathan Slack
GENIUS Andrew Robinson
GEOGRAPHY John Matthews and David Herbert
GEOPOLITICS Klaus Dodds

GERMAN LITERATURE Nicholas Boyle
GERMAN PHILOSOPHY Andrew Bowie
GLOBAL CATASTROPHES Bill McGuire
GLOBAL ECONOMIC HISTORY Robert C. Allen
GLOBALIZATION Manfred Steger
GOD John Bowker
GOETHE Ritchie Robertson
THE GOTHIC Nick Groom
GOVERNANCE Mark Bevir
GRAVITY Timothy Clifton
THE GREAT DEPRESSION AND THE NEW DEAL Eric Rauchway
HABERMAS James Gordon Finlayson
HAPPINESS Daniel M. Haybron
THE HARLEM RENAISSANCE Cheryl A. Wall
THE HEBREW BIBLE AS LITERATURE Tod Linafelt
HEGEL Peter Singer
HEIDEGGER Michael Inwood
HERMENEUTICS Jens Zimmermann
HERODOTUS Jennifer T. Roberts
HIEROGLYPHS Penelope Wilson
HINDUISM Kim Knott
HISTORY John H. Arnold
THE HISTORY OF ASTRONOMY Michael Hoskin
THE HISTORY OF CHEMISTRY William H. Brock
THE HISTORY OF LIFE Michael Benton
THE HISTORY OF MATHEMATICS Jacqueline Stedall
THE HISTORY OF MEDICINE William Bynum
THE HISTORY OF TIME Leofranc Holford-Strevens
HIV AND AIDS Alan Whiteside
HOBBES Richard Tuck
HOLLYWOOD Peter Decherney
HOME Michael Allen Fox
HORMONES Martin Luck
HUMAN ANATOMY Leslie Klenerman
HUMAN EVOLUTION Bernard Wood
HUMAN RIGHTS Andrew Clapham
HUMANISM Stephen Law
HUME A. J. Ayer
HUMOUR Noël Carroll
THE ICE AGE Jamie Woodward
IDEOLOGY Michael Freeden
INDIAN CINEMA Ashish Rajadhyaksha
INDIAN PHILOSOPHY Sue Hamilton
THE INDUSTRIAL REVOLUTION Robert C. Allen
INFECTIOUS DISEASE Marta L. Wayne and Benjamin M. Bolker
INFORMATION Luciano Floridi
INNOVATION Mark Dodgson and David Gann
INTELLIGENCE Ian J. Deary
INTELLECTUAL PROPERTY Siva Vaidhyanathan
INTERNATIONAL LAW Vaughan Lowe
INTERNATIONAL MIGRATION Khalid Koser
INTERNATIONAL RELATIONS Paul Wilkinson
INTERNATIONAL SECURITY Christopher S. Browning
IRAN Ali M. Ansari
ISLAM Malise Ruthven
ISLAMIC HISTORY Adam Silverstein
ISOTOPES Rob Ellam
ITALIAN LITERATURE Peter Hainsworth and David Robey
JESUS Richard Bauckham
JOURNALISM Ian Hargreaves
JUDAISM Norman Solomon
JUNG Anthony Stevens
KABBALAH Joseph Dan
KAFKA Ritchie Robertson
KANT Roger Scruton
KEYNES Robert Skidelsky
KIERKEGAARD Patrick Gardiner
KNOWLEDGE Jennifer Nagel
THE KORAN Michael Cook
LANDSCAPE ARCHITECTURE Ian H. Thompson
LANDSCAPES AND GEOMORPHOLOGY Andrew Goudie and Heather Viles
LANGUAGES Stephen R. Anderson
LATE ANTIQUITY Gillian Clark
LAW Raymond Wacks
THE LAWS OF THERMODYNAMICS Peter Atkins
LEADERSHIP Keith Grint
LEARNING Mark Haselgrove
LEIBNIZ Maria Rosa Antognazza
LIBERALISM Michael Freeden

LIGHT Ian Walmsley
LINCOLN Allen C. Guelzo
LINGUISTICS Peter Matthews
LITERARY THEORY Jonathan Culler
LOCKE John Dunn
LOGIC Graham Priest
LOVE Ronald de Sousa
MACHIAVELLI Quentin Skinner
MADNESS Andrew Scull
MAGIC Owen Davies
MAGNA CARTA Nicholas Vincent
MAGNETISM Stephen Blundell
MALTHUS Donald Winch
MANAGEMENT John Hendry
MAO Delia Davin
MARINE BIOLOGY Philip V. Mladenov
THE MARQUIS DE SADE John Phillips
MARTIN LUTHER Scott H. Hendrix
MARTYRDOM Jolyon Mitchell
MARX Peter Singer
MATERIALS Christopher Hall
MATHEMATICS Timothy Gowers
THE MEANING OF LIFE Terry Eagleton
MEASUREMENT David Hand
MEDICAL ETHICS Tony Hope
MEDICAL LAW Charles Foster
MEDIEVAL BRITAIN John Gillingham and Ralph A. Griffiths
MEDIEVAL LITERATURE Elaine Treharne
MEDIEVAL PHILOSOPHY John Marenbon
MEMORY Jonathan K. Foster
METAPHYSICS Stephen Mumford
THE MEXICAN REVOLUTION Alan Knight
MICHAEL FARADAY Frank A. J. L. James
MICROBIOLOGY Nicholas P. Money
MICROECONOMICS Avinash Dixit
MICROSCOPY Terence Allen
THE MIDDLE AGES Miri Rubin
MILITARY JUSTICE Eugene R. Fidell
MINERALS David Vaughan
MODERN ART David Cottington
MODERN CHINA Rana Mitter
MODERN DRAMA Kirsten E. Shepherd-Barr
MODERN FRANCE Vanessa R. Schwartz
MODERN IRELAND Senia Pašeta
MODERN ITALY Anna Cento Bull
MODERN JAPAN Christopher Goto-Jones
MODERN LATIN AMERICAN LITERATURE Roberto González Echevarría
MODERN WAR Richard English
MODERNISM Christopher Butler
MOLECULAR BIOLOGY Aysha Divan and Janice A. Royds
MOLECULES Philip Ball
THE MONGOLS Morris Rossabi
MOONS David A. Rothery
MORMONISM Richard Lyman Bushman
MOUNTAINS Martin F. Price
MUHAMMAD Jonathan A. C. Brown
MULTICULTURALISM Ali Rattansi
MUSIC Nicholas Cook
MYTH Robert A. Segal
THE NAPOLEONIC WARS Mike Rapport
NATIONALISM Steven Grosby
NELSON MANDELA Elleke Boehmer
NEOLIBERALISM Manfred Steger and Ravi Roy
NETWORKS Guido Caldarelli and Michele Catanzaro
THE NEW TESTAMENT Luke Timothy Johnson
THE NEW TESTAMENT AS LITERATURE Kyle Keefer
NEWTON Robert Iliffe
NIETZSCHE Michael Tanner
NINETEENTH-CENTURY BRITAIN Christopher Harvie and H. C. G. Matthew
THE NORMAN CONQUEST George Garnett
NORTH AMERICAN INDIANS Theda Perdue and Michael D. Green
NORTHERN IRELAND Marc Mulholland
NOTHING Frank Close
NUCLEAR PHYSICS Frank Close
NUCLEAR POWER Maxwell Irvine
NUCLEAR WEAPONS Joseph M. Siracusa
NUMBERS Peter M. Higgins
NUTRITION David A. Bender
OBJECTIVITY Stephen Gaukroger

THE OLD TESTAMENT Michael D. Coogan
THE ORCHESTRA D. Kern Holoman
ORGANIZATIONS Mary Jo Hatch
PANDEMICS Christian W. McMillen
PAGANISM Owen Davies
THE PALESTINIAN-ISRAELI CONFLICT Martin Bunton
PARTICLE PHYSICS Frank Close
PAUL E. P. Sanders
PEACE Oliver P. Richmond
PENTECOSTALISM William K. Kay
THE PERIODIC TABLE Eric R. Scerri
PHILOSOPHY Edward Craig
PHILOSOPHY IN THE ISLAMIC WORLD Peter Adamson
PHILOSOPHY OF LAW Raymond Wacks
PHILOSOPHY OF SCIENCE Samir Okasha
PHOTOGRAPHY Steve Edwards
PHYSICAL CHEMISTRY Peter Atkins
PILGRIMAGE Ian Reader
PLAGUE Paul Slack
PLANETS David A. Rothery
PLANTS Timothy Walker
PLATE TECTONICS Peter Molnar
PLATO Julia Annas
POLITICAL PHILOSOPHY David Miller
POLITICS Kenneth Minogue
POPULISM Cas Mudde and Cristóbal Rovira Kaltwasser
POSTCOLONIALISM Robert Young
POSTMODERNISM Christopher Butler
POSTSTRUCTURALISM Catherine Belsey
PREHISTORY Chris Gosden
PRESOCRATIC PHILOSOPHY Catherine Osborne
PRIVACY Raymond Wacks
PROBABILITY John Haigh
PROGRESSIVISM Walter Nugent
PROTESTANTISM Mark A. Noll
PSYCHIATRY Tom Burns
PSYCHOANALYSIS Daniel Pick
PSYCHOLOGY Gillian Butler and Freda McManus
PSYCHOTHERAPY Tom Burns and Eva Burns-Lundgren
PUBLIC ADMINISTRATION Stella Z. Theodoulou and Ravi K. Roy
PUBLIC HEALTH Virginia Berridge
PURITANISM Francis J. Bremer
THE QUAKERS Pink Dandelion
QUANTUM THEORY John Polkinghorne
RACISM Ali Rattansi
RADIOACTIVITY Claudio Tuniz
RASTAFARI Ennis B. Edmonds
THE REAGAN REVOLUTION Gil Troy
REALITY Jan Westerhoff
THE REFORMATION Peter Marshall
RELATIVITY Russell Stannard
RELIGION IN AMERICA Timothy Beal
THE RENAISSANCE Jerry Brotton
RENAISSANCE ART Geraldine A. Johnson
REVOLUTIONS Jack A. Goldstone
RHETORIC Richard Toye
RISK Baruch Fischhoff and John Kadvany
RITUAL Barry Stephenson
RIVERS Nick Middleton
ROBOTICS Alan Winfield
ROCKS Jan Zalasiewicz
ROMAN BRITAIN Peter Salway
THE ROMAN EMPIRE Christopher Kelly
THE ROMAN REPUBLIC David M. Gwynn
ROMANTICISM Michael Ferber
ROUSSEAU Robert Wokler
RUSSELL A. C. Grayling
RUSSIAN HISTORY Geoffrey Hosking
RUSSIAN LITERATURE Catriona Kelly
THE RUSSIAN REVOLUTION S. A. Smith
SAVANNAS Peter A. Furley
SCHIZOPHRENIA Chris Frith and Eve Johnstone
SCHOPENHAUER Christopher Janaway
SCIENCE AND RELIGION Thomas Dixon
SCIENCE FICTION David Seed
THE SCIENTIFIC REVOLUTION Lawrence M. Principe
SCOTLAND Rab Houston
SEXUALITY Véronique Mottier
SHAKESPEARE'S COMEDIES Bart van Es
SIKHISM Eleanor Nesbitt
THE SILK ROAD James A. Millward
SLANG Jonathon Green

SLEEP Steven W. Lockley and Russell G. Foster
SOCIAL AND CULTURAL ANTHROPOLOGY John Monaghan and Peter Just
SOCIAL PSYCHOLOGY Richard J. Crisp
SOCIAL WORK Sally Holland and Jonathan Scourfield
SOCIALISM Michael Newman
SOCIOLINGUISTICS John Edwards
SOCIOLOGY Steve Bruce
SOCRATES C. C. W. Taylor
SOUND Mike Goldsmith
THE SOVIET UNION Stephen Lovell
THE SPANISH CIVIL WAR Helen Graham
SPANISH LITERATURE Jo Labanyi
SPINOZA Roger Scruton
SPIRITUALITY Philip Sheldrake
SPORT Mike Cronin
STARS Andrew King
STATISTICS David J. Hand
STEM CELLS Jonathan Slack
STRUCTURAL ENGINEERING David Blockley
STUART BRITAIN John Morrill
SUPERCONDUCTIVITY Stephen Blundell
SYMMETRY Ian Stewart
TAXATION Stephen Smith
TEETH Peter S. Ungar
TELESCOPES Geoff Cottrell
TERRORISM Charles Townshend
THEATRE Marvin Carlson
THEOLOGY David F. Ford
THOMAS AQUINAS Fergus Kerr
THOUGHT Tim Bayne
TIBETAN BUDDHISM Matthew T. Kapstein
TOCQUEVILLE Harvey C. Mansfield
TRAGEDY Adrian Poole
TRANSLATION Matthew Reynolds
THE TROJAN WAR Eric H. Cline
TRUST Katherine Hawley
THE TUDORS John Guy
TWENTIETH-CENTURY BRITAIN Kenneth O. Morgan
THE UNITED NATIONS Jussi M. Hanhimäki
THE U.S. CONGRESS Donald A. Ritchie
THE U.S. SUPREME COURT Linda Greenhouse
UTOPIANISM Lyman Tower Sargent
THE VIKINGS Julian Richards
VIRUSES Dorothy H. Crawford
VOLTAIRE Nicholas Cronk
WAR AND TECHNOLOGY Alex Roland
WATER John Finney
WEATHER Storm Dunlop
THE WELFARE STATE David Garland
WILLIAM SHAKESPEARE Stanley Wells
WITCHCRAFT Malcolm Gaskill
WITTGENSTEIN A. C. Grayling
WORK Stephen Fineman
WORLD MUSIC Philip Bohlman
THE WORLD TRADE ORGANIZATION Amrita Narlikar
WORLD WAR II Gerhard L. Weinberg
WRITING AND SCRIPT Andrew Robinson
ZIONISM Michael Stanislawski

Available soon:

MILITARY STRATEGY Antulio J. Echevarria
NAVIGATION Jim Bennett
THE HABSBURG EMPIRE Martyn Rady
INFINITY Ian Stewart
CIRCADIAN RHYTHMS Russell Foster and Leon Kreitzman

Tristram D. Wyatt

ANIMAL BEHAVIOUR

A Very Short Introduction

Great Clarendon Street, Oxford, OX2 6DP,
United Kingdom

Oxford University Press is a department of the University of Oxford.
It furthers the University's objective of excellence in research, scholarship,
and education by publishing worldwide. Oxford is a registered trade mark of
Oxford University Press in the UK and in certain other countries

First edition published in 2017

Impression: 1

Published in the United States of America by Oxford University Press
198 Madison Avenue, New York, NY 10016, United States of America

British Library Cataloguing in Publication Data
Data available

Library of Congress Control Number: 2016956624

ISBN 978–0–19–871215–2

Printed in Great Britain by
Ashford Colour Press Ltd, Gosport, Hampshire

Contents

Acknowledgements

Many people have encouraged my interest in animal behaviour. I thank early influences and my colleagues, students, and many others who have advised me over the years. For their helpful comments on the whole manuscript I am especially grateful to Kate Goldenberg, Linda Sims, Robert Taylor, Joan Wyatt and Vivian Wyatt; my editors at Oxford University Press, Latha Menon and Jenny Nugee; and my anonymous readers. For reading chapters and expert advice on particular topics I would like to warmly thank Aaron Allen, Dora Biro, Victoria Braithwaite, Heather Eisthen, Stephen Goodwin, David Haig, and Megan Neville. David Benz, Uli Ernst, Bruce Schulte, Jim Thorne, and Tobias Uller kindly advised me at the scoping stage.

List of illustrations

6 Fruitfly courtship involves many steps. The male fruitfly sings to the female by vibrating his extended wing towards her. If a duet of chemical signals next confirms that they are of the same species, he licks her genitalia and attempts to copulate, but the choice is hers **22**

Fig. 9.5 from B. Burnet and K. Connolly (1974). Activity and sexual behaviour in *Drosophila melanogaster*. In J.H.F. van Abeelen (ed.), *The genetics of behaviour*. North Holland, Oxford, pp. 201–58. Copyright Elsevier 1974

7 In vertebrates, internal and external influences are integrated in the brain's hypothalamus (Hypo), which sends hormone signals stimulating the pituitary gland (Pit) to release its hormones. These activate peripheral hormone-producing (E) tissues, such as the ovaries, testes, and adrenal glands **25**

Based on Fig. 1, p. 179, reprinted from *Frontiers in Neuroendocrinology*, Vol. 36, Robert S. Bridges, Neuroendocrine regulation of maternal behavior, pp. 178–96. Copyright 2015, with permission from Elsevier

8 A jewel wasp (*Ampulex compressa*) female delivers precision doses of neurotoxins as it stings the head of the cockroach (*Periplaneta americana*). The wasp's larva will feed on the living but immobilized host **27**

Figure kindly provided by Prof. Frederic Libersat, Dept of Life Sciences and Zlotowski, Center for Neurosciences, Ben Gurion University

9 A sketch of Konrad Lorenz attracting his imprinted goslings, made by Niko Tinbergen on a visit to Lorenz in Austria **32**

Figure kindly provided by the Tinbergen estate

10 The development of an individual's behaviour during its lifetime involves a complex interaction between its genes and its physical and social environment (including other species) **37**

Based on Fig. 1 in *Apidologie*, A.G. Dolezal and A.L. Tóth, Honey bee sociogenomics: a genome-scale perspective on bee social behavior and health, Vol. 45 (2013), pp. 1–21, Apidologie © INRA, DIB, and Springer-Verlag France, 2013. With permission of Springer

11 Sparrows which can hear their own species' songs during development produce normal songs themselves as adults, visualized in sonographs (top). Isolated birds which do not hear their species' song during their sensitive period produce abnormal songs (bottom) **43**

From Fig. 3, A comparison of songs developed in nature and in social isolation by two sparrows, the Song Sparrow (*Melospiza melodia*) and the Swamp Sparrow (*Melospiza georgiana*), in Innateness and the instinct to learn by Peter Marler,

19 Yao honey-hunter Orlando Yassene holds a wild greater honeyguide (temporarily captured for research) in the Niassa National Reserve, Mozambique **73**

Photo kindly provided by Dr Claire N. Spottiswoode

20 Animals balance risk and reward. Sticklebacks usually choose to attack high-density areas of prey, but after a model kingfisher was flown over the tank they chose low-density areas **79**

Republished with permission of John Wiley and Sons Inc, from Figure 3.7, p. 65 in J.R. Krebs and N.B. Davies (1987). *An introduction to behavioural ecology*, 2nd edn. Oxford, Blackwell Scientific Publications; permission conveyed through Copyright Clearance Center, Inc. Data reprinted by permission from Macmillan Publishers Ltd: M. Milinski and R. Heller, Influence of a predator on the optimal foraging behaviour of sticklebacks (*Gasterosteus aculeatus L.*). *Nature*, Vol. 275, pp. 642–4. Copyright (1978)

21 Great tits in woods near Oxford, England, have an average clutch of about 8–9 eggs (bars) (white arrow). When eggs were added or removed in a manipulation experiment, it was discovered that a pair of great tits could rear a larger number successfully (shown by recaptures, dots and curve, black arrow) **80**

Republished with permission of John Wiley and Sons from Fig. 1.5b, p. 13 in N.B. Davies, J.R. Krebs, and S.A. West (2012). *An introduction to behavioural ecology*, 4th edn. Chichester, Wiley-Blackwell; permission conveyed through Copyright Clearance Center, Inc. Data republished with permission of John Wiley and Sons Inc, from C.M. Perrins (1965). Population fluctuations and clutch-size in the great tit, *Parus major L. Journal of Animal Ecology*, Vol. 34, pp. 601–47

22 Male widowbirds with longer tails are more attractive to females. Before the experimental manipulations, all the males were equally attractive (a). After some males had their tails shortened or lengthened (b), females nested most in the territories of males with the super-elongated tails **87**

Reprinted by permission from Macmillan Publishers Ltd: M. Andersson, Female choice selects for extreme tail length in a widowbird, *Nature*, Vol. 299, pp. 818–20. Copyright (1982)

23 Naked mole rats are eusocial mammals with a queen and workers, analogous to social insects. Here workers are shown digging a tunnel together **93**

Fig. 6.11b, page 321 , Manning, A & Dawkins, MS (2012). *An Introduction to Animal Behaviour*, 6 edn. Cambridge, Cambridge University Press, from a photo by David Curl, drawing by Priscilla Barrett. Provided with kind permission from Priscilla Barrett

24 A murmuration of thousands of starlings **98**

Photo by Adam <https://www.flickr.com/photos/aaddaamn/5196833268//>

The publisher and author apologize for any errors or omissions in the above list. If contacted they will be pleased to rectify these at the earliest opportunity.

Chapter 1
How animals behave (and why)

A hummingbird hovering in front of a flower to collect nectar; an octopus changing its colour for camouflage on a coral reef; a dragonfly motionless on a leaf, basking in the sun. All of these are examples of animal behaviour. Behaviour is a key way in which animals interact with their world. It is how animals find and choose mates, look after their young, find food, avoid becoming food for predators themselves, and build nests and burrows.

Television documentaries on animal behaviour are popular worldwide. The best of these programmes are based on a deep understanding of animal behaviour gained through years of hard work by scientists, graduate students, and volunteers. Meerkats, small group-living mongooses in southern Africa, have been so extensively studied and filmed that we can follow individuals through their lives like characters in an animal soap opera. The Kalahari Desert meerkats, *Suricata suricatta*, have been followed over generations (Figure 1). They are so habituated to humans that they will climb on and off weighing scales when a scientist wants to weigh an animal. It is remarkable that behaviour which at one time could only be observed by dedicated field workers is now readily available for all of us to see. While you are reading this book, from time to time I hope you will search the web for the behaviours and animals that you are reading about. There is a good chance you will find relevant films to watch.

From hunting to ethology

Our current interest in animal behaviour follows in a long tradition tracing back to the beginnings of recorded history. Humans have probably always been keen observers of animal behaviour. For example, 30,000-year-old prehistoric cave paintings in the Dordogne of southern France show hunting scenes with deer, bison, horses, and many other recognizable mammals. One painting shows a bull sniffing a cow, just as we might draw it now. Paintings of birds, fish, and mammals decorate the royal tombs from ancient Egypt some 5,000 years ago. Some early observers were surprisingly accurate: modern studies have confirmed many of Aristotle's observations in the 4th century BCE about European cuckoos, parental care by a catfish, and the way the nightingale learns its song.

In the 19th century, descriptions of animal behaviour tended to be anecdotal and sentimentally anthropomorphic (with human emotions such as remorse attributed to animals). Charles Darwin's chapter on instinct in *The Origin of Species* (1859) was an exception. Darwin described phenomena such as cuckoos laying their eggs in host birds' nests and experiments on the behaviour of honeybees constructing the hexagonal cells of honeycomb. He was interested in how natural selection might have shaped rudimentary behaviours into sophisticated instincts, a question which continues to interest scientists today. In later books, including *The Descent of Man, and Selection in Relation to Sex* (1871), Darwin extended his ideas about animal behaviour to human evolution. He suggested, for example, that the geographical differences in our facial features could be largely the result of female choice.

The development of more scientific and objective study of animal behaviour in late 19th and early 20th century owes much to pioneering scientists including Margaret Washburn who wrote

1. A group of meerkats standing vigilant, scanning for predators, in the Kalahari Desert, South Africa.

a highly influential textbook, *The Animal Mind* (1908), the African-American scientist Charles H. Turner, the Russian scientist Ivan Pavlov, John Watson, and B.F. Skinner. Their 'behaviourist' approach, using laboratory experiments to explore how animals behave (in particular how they learn), became the scientific discipline we now call *experimental psychology*.

A second movement, which became the discipline of animal behaviour that we recognize today, left the laboratory behind and instead focused on studying animals in the wild. The evolutionary origins of behaviour became the unifying question. Karl von Frisch, Konrad Lorenz, and Niko Tinbergen shared a Nobel Prize in 1973 for their roles in helping to create this new discipline which they called *ethology*. Von Frisch studied the visual world of the honeybee and discovered the dances that bees use to communicate the location of flowers offering good nectar. Lorenz's work on courtship in different duck species suggested how such behaviours might have evolved. Tinbergen carried out pioneering experiments on the function and evolution of the behaviours of many animals, including digger wasps, stickleback fish, and birds.

The early ethologists focused on species-level descriptions of behaviour (for example, describing the typical courtship behaviours of males and females of a species). The next revolution in the study of animal behaviour came with the recognition that selection acts at the level of genes in individuals, notably advocated by the American evolutionary biologist George C. Williams in his 1966 book, *Adaptation and Natural Selection*. Evidence came, for example, from David Lack's findings that breeding pairs of birds such as great tits maximize the number of young they can produce, rather than limiting the number of young 'for the good of the species' as had been commonly thought by many biologists.

The popular breakthrough came with Richard Dawkins' book, *The Selfish Gene*, in 1976. Dawkins showed how looking at natural selection from the 'gene's-eye-view' could explain how behaviours which benefit the holder of the gene, but not the species as a whole, can evolve. The influence of the book was not limited to the general public. In a rare example of a popular science book helping change a whole discipline, *The Selfish Gene* was also enormously influential on the thinking of professional biologists. The idea of 'for the good of the species' started to fall out of favour.

When looking at behaviour from a gene's-eye-view, individual animals can be regarded as 'survival machines' by which genes survive to replicate, leaving more copies of themselves in the next generation. For example, parent birds that lay the optimal number of eggs, which leads to the greatest number of surviving offspring, will leave more individuals in subsequent generations, with copies of the optimal-egg-number-to-lay gene(s). When we talk about a 'gene for' a particular structure or behaviour, such as egg number, we do not mean that the trait is due to that gene alone. Rather, many interacting genes are likely to be involved in producing the trait. Nonetheless, if a mutation in a particular gene leads to a difference in behaviour, as a shorthand we could say it was the 'gene for' the behaviour (Chapter 3).

Animals can also pass on copies of their genes to the next generation by helping relatives, which share those genes, to survive or reproduce. This is the basis of inclusive fitness or kin selection, proposed by W.D. Hamilton in groundbreaking papers in the 1960s and brought to a wider audience by Richard Dawkins in *The Selfish Gene*. E.O. Wilson's 1975 monumental book *Sociobiology* was also highly influential.

Meerkats offer a good example of how biologists have shifted their focus when studying behaviour. In the 1960s, scientists described the behaviour of meerkat groups in general terms, focusing on the number of animals in each group and descriptions of the ways that subordinate meerkats helped their parents to rear their siblings. Now, the focus has shifted to the different behaviours of individuals within a group and how behaviours such as helping might have evolved. For example, babysitting young siblings may be expensive to the helper in lost feeding time, so what are the individual benefits and costs to being a helper? These kinds of questions underlie the approach now known as *behavioural ecology*, which combines ethology and ecology (Chapter 6).

Meerkats also offer an example of the way that behavioural studies tend to be concentrated on larger, attractive, charismatic terrestrial animals such as mammals and birds. This is in part because these animals may be easier to observe. If we had been aquatic ourselves we might have carried out more studies of fish and marine invertebrates such as crabs. Our views of parental care might have been different (if fish look after their young, it is usually the male that looks after them).

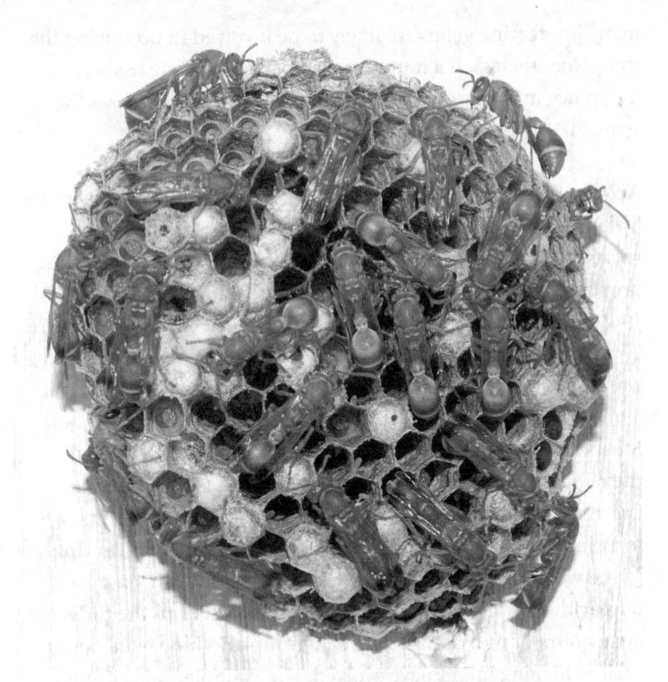

2. Adults of the Indian paper wasp, *Ropalidia marginata*, interact on the surface of their nest, making them ideal for behavioural observations. The nest has some capped brood (*pupae*) which will later emerge as adults.

Animals found near universities in Europe and North America also tend to be over-represented, meerkats notwithstanding. For example, the great tits in Wytham woods, just outside Oxford, are heavily studied. A nice counter-example of an invertebrate in the tropics comes from the detailed studies of the Indian paper wasp, *Ropalidia marginata*, pioneered by Raghavendra Gadagkar in Bangalore (Figure 2). Many key insights into the evolution of social insects (the bees, ants, and wasps) have come from his long-term research focus on this paper wasp, which forms small colonies with a queen who looks like her fellow workers. Gadagkar points out that behavioural research need not be expensive, and good work can be done anywhere: interesting animals occur everywhere. It is also possible to make a contribution as an amateur; in 'Further reading' you will find suggestions of ways to get involved with *citizen science*.

Tinbergen's four questions

Behaviours can be described in many different ways. Niko Tinbergen made a key contribution to the development of the discipline by observing that biologists could examine the same behaviour from different perspectives. These complementary perspectives can be summarized as four questions, which can be asked about any feature of an animal, not just a behaviour: What is it for? How did it develop during the lifetime of the individual? What mechanisms control the behaviour? How did it evolve over evolutionary time?

What is the behaviour for? First, how does a behaviour help an animal survive or help improve its success in reproduction? Tinbergen's research group explored this question in field experiments. For example, they found that the removal of the conspicuous empty egg shells from the nest by the parent black-headed gull, *Larus ridibundus*, reduced the chance of the newly hatched nestlings being discovered by visual hunters such as carrion crows and herring gulls (Figure 3).

How does the behaviour develop in an individual? As the young animal develops from the egg to adult, behaviours develop under the influence of genes and environment. For example, the amount of licking and grooming of pups by the mother mouse will affect the pups' later behaviour including how they parent when adult themselves. Animals may have sensitive periods when they are especially primed to learn. For example, young songbirds must hear adult song in order to be able to sing an attractive song when adult (Chapter 3). Social learning explains how techniques for finding or treating food such as washing sand off sweet potatoes

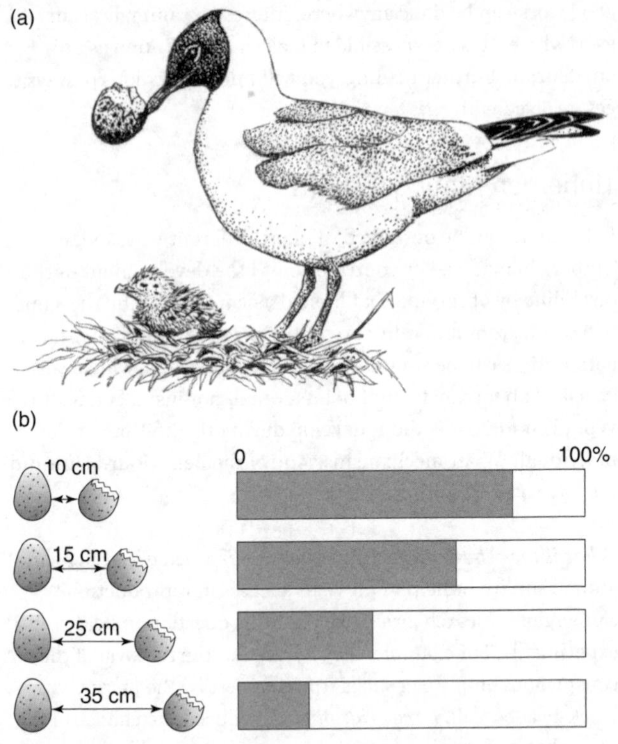

3. (a) Black-headed gull parents remove the empty egg shell after the chick hatches; (b) Tinbergen tested the hypothesis that the white, jagged edges give predators visual clues. In field experiments, the closer the empty egg shell was to a nest, the more whole eggs were found by predators.

by Japanese macaque monkeys were passed on from generation to generation (Chapter 4).

What mechanisms control the behaviour? This question explores the factors, internal and external, which cause a behaviour to occur at a particular moment. These factors could be investigated at many levels, from a molecular focus to a whole animal view. For example, we now have a detailed understanding of the mechanisms involved in how birds learn and produce their songs, from which genes are active in the brain during development, how neural circuitry links the different parts of the brain, to the role of sensory inputs such as the songs of other birds. Which behaviours happen when is also influenced by hormones released in response to interactions with other animals, and factors such as day length or season.

How did the behaviour evolve? We can look at how behaviours evolve just as we can look at the way body parts change over millions of years of evolutionary time, for example the way that bat wings and whale flippers have both evolved from the limbs of a common ancestor. (Features or traits shared by common ancestry are called *homologous*.) Evolution frequently works by co-opting existing features for a new function. For example, when a predator comes to the mouth of its burrow, the burrowing owl (*Athene cunicularia*) mimics the sound of a rattlesnake's rattle. This persuades a mammalian predator that there is a dangerous snake in the burrow rather than a harmless owl. It is an odd sound for an owl to make. It seems to have evolved from a food begging call, which is still used for this function by the young of related owl species.

Behaviours can evolve quickly. Over ten years, male field crickets (*Teleogryllus oceanicus*) on the Hawaiian island of Kauai fell silent. By 2001, only a single male was heard and crickets were hard to find. In 2003, the crickets were back in numbers but their songs were silent. The cause of the silence had been the arrival on the island of a deadly parasitoid fly (*Ormia ochracea*) which uses a male cricket's song to find him, parasitizing him with her maggots, which then eat him alive. Only silent male crickets escaped the fly.

Scientist Marlene Zuk discovered that the surviving silent cricket males had a previously rare mutant form of wing which lacked the 'teeth' and 'scraper' used to make the sound. Under the deadly selection pressure from the fly, the mutant silent males had gone from rare to common in just twenty generations. On another

Hawaiian island, Oahu, the males also fell silent two years later with a seemingly similar silent mutant, again selected for by the arrival of the parasitoid fly. However, the mutations that produce the silent wing on Oahu have arisen independently. This is a nice example of convergent evolution in which silent males evolved independently and in different ways on the two islands.

At the other end of timescales, fossils sometimes preserve the behaviour of animals millions of years ago. For example, a Permian shark that was fossilized having just eaten salamanders, which had themselves just eaten small fish, showed an early food web all in a single fossil and its stomach from 300 million years ago. On the other hand, there are other animals we know only from the fossilized traces of feeding movements and burrows they left in the late Precambrian mud, as their soft bodies were not preserved. Signs of parental care in dinosaurs, with adults closely associated with nests of eggs, have been found for many kinds of dinosaurs. Biomechanical models based on animals living today allowed palaeontologists to use measurements of trails of dinosaur foot prints and muscle traces on fossilized bones to show that the dinosaur *Tyrannosaurus rex* could probably have run at a still scary 20 kilometres per hour (km/h) rather than the 70 km/h sometimes claimed. More controversially, it seems that the colours of dinosaurs and other animals can also be deduced, from microscopic structures in their fossilized scales and feathers. For example, it has been suggested that a bird-like dinosaur *Anchiornis huxleyi* had a crest of red feathers on its head. Colour patterns like this lead to speculation about courtship and other behaviours the dinosaurs might have shown.

Integrating the four questions

The ways that newly born mammals and their mothers have co-evolved to ensure a successful first milk meal provide good examples of how Tinbergen's four questions can be used to understand a single phenomenon, in this case the importance of a newborn mammal's ability to find and grasp a nipple as speedily as possible.

Function. The function of the nipple-search behaviour is clear: a prompt first meal is essential for survival. In all mammals studied so far, including humans, neonatal survival is strongly influenced by the first milk meals. Up to half of human babies do not get an optimal milk intake on the first day and, without intervention, this can affect survival. In a study of home-born infants in rural West Africa, an hour's post-birth delay in the first milk meal could explain more than a fifth of neonatal mortality in the first year. The special importance of the first milk meal comes in part from the colostrum it contains, giving antibodies offering passive immunization, an inoculation of beneficial gut bacteria, and growth factors. Hormones in the milk affect the baby's behaviours such as sleep and learning.

Development. How does the nipple finding and suckling behaviour develop in the life of the young mammal? The searching behaviour itself does not seem to be learned: the newborn mammal shows a stereotypical search (human babies placed on their mother's breasts do this). Newborn mammals respond to odour and other stimuli such as warmth and shape that indicate the location of the nipple. All mother rabbits (*Oryctolagus cuniculus*) produce the same mammary pheromone molecule in their milk, which rabbit pups respond to without the need for learning. The rabbit pups' response to the pheromone is greatest in the days after birth and then successively drops as the pups' eyes open and then drops further with weaning (moving to solid food), and after the third week there is almost no response. For the first four days after birth, the rabbit mammary pheromone also causes the pup to learn any other smell associated with it so the pup also learns the individual smell of its mother as a positive cue to find the nipples. In addition to their innate responses, all mammal newborns are learning machines.

Mechanisms. An important controlling mechanism in the production of milk for the mother rabbit, and most mammals including humans, is the tactile stimulus to the nipple from the sucking behaviour of the infant. Nerves stimulated in the nipple send signals to the *hypothalamus* in the brain, causing release of the hormone *prolactin*, which stimulates milk production, and the hormone *oxytocin*, which stimulates milk ejection. These hormones may underlie nursing behaviour and maternal motivation (and in some species, bonding). In some species, including humans, by inhibiting ovarian activity the hormone prolactin may help delay the next conception until after weaning of the current offspring.

Evolution. The stimuli evolved by female mammals to enable their offspring to find a nipple include odours and warmth, detectable using the senses that all newborn mammals have. The needs of the newborns, which may be blind and helpless, have also selected for the evolution of easily detectable, graspable nipples or teats.

Tinbergen's four questions are still relevant to researchers investigating animal behaviour today. Which types of question get most attention from a particular generation of biologists change according to fashions (scientists are just as guilty of this as other people). Sometimes the types of questions that can be answered expand, perhaps because a new technique such as DNA-fingerprinting makes new kinds of enquiry possible. However, we still need good observation, good questions, thoughtful experiments, and a feeling for the organism, which the early behavioural biologists like Niko Tinbergen used to such great effect.

Chapter 2
Sensing and responding

How an animal behaves is coordinated by nerves and hormones, in their different and complementary ways. Stimuli, such as the sound of a predator, cause fast behavioural responses coordinated by nerve signals. The stimuli also cause longer lasting physiological changes via hormones which release energy sources that will be needed for the muscle action necessary for escape.

Bats and moths

By flying at night, moths avoid the danger of being eaten by birds, but instead they risk being caught by bats, the specialist aerial night-time predators. In the dark, bats use ultrasonic sonar to 'see' the world, hearing the echoes reflected back from the landscape and their prey. The loud sonar signal produced by the bat is like a searchlight scanning the sky ahead, getting louder in the final stages of attack as the bat homes in on the moth (Figure 4). Moths, like many other insects, have independently evolved ears that can detect bats' sonar beams and prompt fast evasive action. The late Kenneth Roeder, a pioneer of studies of bats and moths, suggested jangling your keys under a lamppost on a summer night to see moths dropping out of the air as they respond to the bat-frequency ultrasonic sounds made by the jangling keys. Do try it.

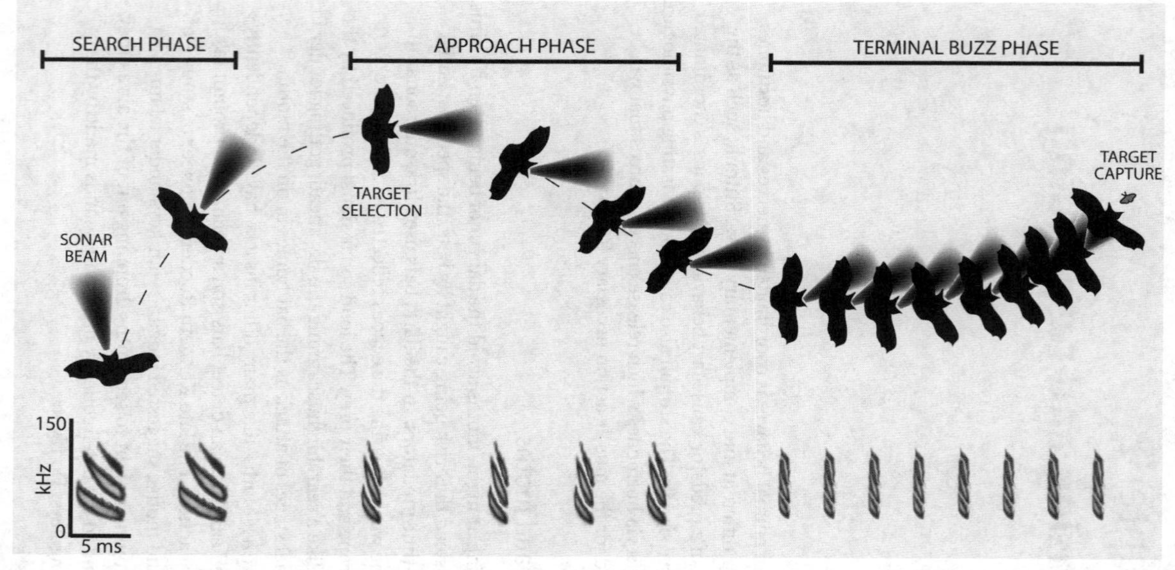

4. A hunting bat's ultrasonic calls change, as it targets an insect, with shorter calls, wider band width, and faster repetition. Moths can hear the bat's calls and take evasive action.

Both animals in the aerial dogfight need to act fast, using behaviours enabled by nerve cells (neurons). Moths provide a useful model system for studying how a stimulus, the bat call, is detected and leads to evasive behaviour. The moth has two ears, one on each side of the *thorax* (the moth's middle section that carries the wings). The drum-like ear facing the bat vibrates most as the sound waves from the bat call hit the moth. This vibration stretches a specialized sensory neuron, causing an electrical signal called an *action potential* to travel down the neuron's *axon*, or 'cable'. The signal is then relayed to other neurons and interneurons, continuing to the flight control nerve centres in the thorax and up into the brain. Ultimately, these nerve signals change wing movements on the side the sound came from, resulting in the moth's evasive manoeuvre. The ear sends two signals to the brain, first from the softer searching calls of a cruising bat 30 metres away, which sends the moth turning and flying away before the bat detects the moth. Second, if the ear detects the loud buzz calls as the bat homes in, it triggers the moth to make a last-second loop or power dive to safety. In the sixty-five-million-year evolutionary arms race between moths and bats, as moths evolved ears some bats have switched to ultrasonic frequencies not detected by the moths. Some moth species counter with sonar-jamming chirps or ultrasonic sounds to startle the bat as it gets close.

Sensory inputs

Out of all the information that could be sensed, animals evolve to detect only what affects their survival or reproductive success. The first stage of selectivity is the sensitivity of their sense organs. The moth ear is a good example of selective sensitivity: it only responds to the ultrasound frequencies produced by bat species that hunt them. Other sounds unimportant to the moth are not detected. Selective sensitivity is a common feature of sense organs.

Animals have evolved sensitivity to a wide range of different kinds of stimuli, from the familiar vision, hearing, smell, and taste to others which we do not experience ourselves, such as sensitivity to polarized light, electrical fields, and magnetic fields. Each kind of stimulus needs a different kind of sensor. Using a technology analogy, this is similar to the way we can use different sensory devices plugged into a computer: a microphone, a webcam camera, a temperature sensor, and a weight sensor. What each of the plug-in sensors has in common is that they have a particular *transducer*, a component which transduces (or converts) one kind of energy, such as sound, into an electrical signal that can be interpreted by appropriate software in the computer. Animal sensory systems offer a similar pattern: for each kind of stimulus the transducer is a sensory neuron with a particular sensitivity (for example, a sensitivity to light, or a sensitivity to movement or stretch). This neuron transducer converts the external stimulus—a sound wave, a photon, or an odorant molecule—into electrical action potentials sent to neurons in the brain. The action potentials are the same whatever the sensor—it is the wiring in the brain that interprets the incoming signal as sound or a visual image, depending on the sensor it is coming from.

The specificity of the received signal comes from the specificity of the transducer. For example, a sensory neuron in the eye will be sensitive to a range of light wavelengths depending on the type of light-sensitive molecules (opsins) it has. Different kinds of animals have slightly different opsin molecules and hence are sensitive to different wavelengths. For example, unlike humans, honeybee eyes do not detect what we call 'red' but, like many birds, they can see ultraviolet light (so named because it is beyond human senses, hence 'ultra').

That bats might use sounds, beyond our hearing, to enable them to fly and hunt in complete darkness was long suspected. However, confirmation only came in the late 1930s when students Donald Griffin and Robert Galambos used early microphones and electronics that could convert the bat's ultrasonic signals into the lower sound frequencies we can hear (or can visualize in other instruments). Their work helped establish the field of *neuroethology* which explores the sensory worlds of animals and the neural basis of animal behaviour. A similar story of discovery exists for each of the senses. For example, the extraordinary sensory world of electric fish was only revealed when we had instruments to measure and create electric fields.

The night-flying male moths are not using sounds to find the female moths. Instead they are using the smells that the females give off. These smells are chemical signals called *pheromones*. Male moths have antennae covered with thousands of sensory 'hairs'. Each hair contains neurons with smell (*olfactory*) receptors which are stimulated by molecules in the sex pheromone blend produced by the female of their species. Each neuron has just one type of receptor, sensitive to one molecule of the female's pheromone blend (every moth species has a different blend of molecules). When the right molecule stimulates the receptor, it sets off the action potential passing down the neuron's axon to the brain. All the thousands of neurons with the same receptor send their axons to the same brain area, a *glomerulus* (named after a ball of wool), in an enlarged part of the moth's brain. When the correct combination of stimulated glomeruli indicates that the signal comes from a female of his species, the male moth's brain sends signals to the wing motor muscles and he flies upwind. (Incidentally, the male moth is not flying up a concentration gradient of pheromone as was once thought. Rather, he zig-zags his way, stimulated to fly upwind each time his antennae hit a pocket of air that has passed over the female and picked up her pheromone.) Moth pheromones have offered remarkable opportunities to study the evolution of animal communication during *speciation* (the evolutionary process whereby new species arise). We can study the genes for receiving the signal in the male: each olfactory receptor is a protein coded for by a gene. We can also investigate the genes coding for the enzymes that produce the molecules that form the female's pheromone signals. As new species form we can trace the changes in both signal and receiver.

Like the male moth, other animals devote large proportions of their brain to processing sources of information important to them. The North American star-nosed mole (*Condylura cristata*) has a spectacular pink nose with a star of twenty-two rays (Figure 5), which it moves in a flurry of motion as it travels quickly along its burrows. The rays are covered with 1,000 touch-sensitive organs which send their signals to the *somatosensory* (or 'touch') *cortex* in the brain. The part of the cortex receiving inputs from the nose-star has twenty-two sections, one for each of the rays. More than 25 per cent of the somatosensory cortex is devoted to signals from the pair of rays closest to its mouth. If a potential prey such as a small insect is touched by one of the other rays, the mole swings its head to assess it with this pair of rays (Figure 5). The wide sweep of the rays coupled with fast movement allow the star-nosed mole to have the fastest food handling of any mammal. It is able to touch, seize, and process small prey in just 120 milliseconds.

5. As it dashes along its burrow, a star-nosed mole uses the twenty-two rays around its nose to sense prey items by touch. The pair of rays closest to its mouth (arrowed) are especially sensitive.

Neural circuits

Early neuroethologists had to treat animals as 'black boxes'. Starting with model animals chosen for their size and ease of handling, such as sea slugs, and moths and other insects, neuroscientists were able to work out simple neural circuits underlying behaviours such as learning and movement. One of the appeals of the moth's ear for experimenters in the 1950s was that there were only two sensory neurons, compared with the hundreds of thousands coming from a vertebrate's eye or ear. In addition, unlike mammals, in many invertebrates each neuron in a *ganglion* (or concentration of neurons) can be identified, named, and its connections followed. The same neurons can be found and studied in other individuals. Another enabling breakthrough was the increasing sophistication of techniques using electrodes to record electrical impulses from nerves, visualizing the electrical spikes of action potentials as a trace on an *oscilloscope* (such as a hospital heart monitor screen). Working with model animals and simple circuits is valuable because what we learn can help us understand more complicated behaviours in other animals.

Studies of learning in sea slugs explored changes in a simple defensive reflex response. Initially the sea slug vigorously retracted its gills when its body was gently touched. The response diminished with repeated touches: a phenomenon called *habituation* (a kind of learning). Conversely, a nasty stimulus such as a pinch to the animal's tail enhanced the gill retraction response, an effect called *sensitization*, which can be long lasting. The changes in behaviour were discovered to be due to changes in the communicating connections (*synapses*—the specialized points of contact between neurons) between the sensory neurons detecting the touch and the motor neurons triggering the muscles to retract the gills. The neuron passing on the signal releases molecules (*neurotransmitters*) into the *synaptic gap*. These molecules stimulate the neuron on the other side. In the sea slug habituation example, the sensory neurons released less and less neurotransmitter with each gentle touch to the body. In the sensitization example, more neurotransmitter was released but, perhaps more importantly, the receiving neuron showed an increased sensitivity to the neurotransmitter. We now know that the synaptic phenomena explored in sea slugs and other invertebrates underlie the learning processes of vertebrates.

Flying locusts beat their wings twenty times a second. What sets the beat? In the flying locust, the wing muscles are controlled by coordinated rhythmic signals from a neural circuit. Surprisingly, once activated, the neural circuit generates the rhythm by itself, even if taken out of the locust. The circuit is a *central pattern generator*, a phenomenon first investigated by the German neuroethologist Erich von Holst. Central pattern generators have been found in animals of all kinds. They produce rhythmic motor patterns such as walking in cockroaches; feeding in lobsters; courtship song in fruitflies; swimming in sea slugs, fish, and tadpoles; and walking and breathing in mammals.

Animals usually carry out a limited number of behaviours at a time. Neurons in command centres in the brain are one way that animals prioritize behaviours in an adaptive way. Command centres interact to stimulate or inhibit each other in complicated ways so that, for example, the right pattern generator(s) are activated. Male moths flying at night in search of females need to balance the chance to reproduce against the risks of being eaten by bats. Male moths flying upwind in response to a strong pheromone plume from a female ignore bat calls until the bat is very close, and only then will they power dive or drop from the sky. Males that take this risk will produce more offspring in their lifetime, on average.

While some behaviours are a response to an external stimulus such as a pheromone, much of any animal's behaviour is generated by internal neural processes—animals are not merely input–output robots. Even in the tiniest brains, such as those of flies or nematode worms, for example, behavioural flexibility seems to be the evolved norm, and nervous systems are characterized by constantly changing activity. A fruitfly in a constant environment generates its own spontaneous turns and movements, seemingly probing the environment for feedback to learn from. In a nematode worm, an investigation into searching behaviour revealed a sensory neuron which sends a clear on/off signal to a motor neuron in a simple circuit. However, the circuit also includes two other interneurons which seem to introduce variability so that the resulting behaviour is unpredictable. Unpredictability is an advantage in interactions with competitors and predators. Prey with predictable behaviour get eaten easily. The escape response of cockroaches offers a good example. If cockroaches always shot off at a predictable angle from an approaching predator, they would be easy to catch. Instead, when experimenters watched hundreds of escapes by cockroaches, they observed that the common escape directions covered a range of angles, which were very hard for the predator to learn.

Neural circuits for complex behaviours are extremely hard to study, particularly in vertebrates with brains containing billions of neurons. Flies are a bit easier, with brains containing 100,000 neurons—though still immensely challenging to study, and even with flies we are just at the very beginning of an understanding that is likely to remain incomplete for decades to come.

Nonetheless, the courtship behaviour of the fruitfly *Drosophila melanogaster* provides an important model system. Normally the courtship follows a stereotyped sequence of male and female responses (Figure 6). However, a mutation in a gene called *fruitless* leads to 'fruitless' courtship because the males court each other, forming lines of dancing males. While there is undoubtedly more to the story, the gene *fruitless* turns out to be crucial in the development of the neural circuit involved in male courtship behaviour. About 2,000 neurons carrying the male version of

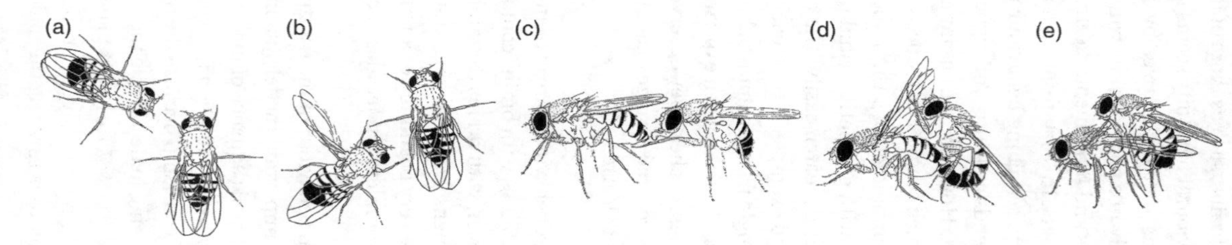

6. Fruitfly courtship involves many steps. The male fruitfly sings to the female by vibrating his extended wing towards her. If a duet of chemical signals next confirms that they are of the same species, he licks her genitalia and attempts to copulate, but the choice is hers.

fruitless form the male neural circuit. These include the male sensory neurons and interneurons that detect and process female pheromones, and motor neurons that regulate the male's wing vibration song. A small group of male-specific central neurons integrates the courtship inputs from the fly's smell, taste, visual, and hearing systems.

New molecular tools are enabling the study of more complex neural circuits in model animals such as the fruitfly, zebrafish, and mouse. *Optogenetics* is one of these tools, allowing a neural circuit to be switched on by the light from a laser beam directed to particular neurons that have been genetically engineered to have light-activated proteins such as channelrhodopsin. The researcher can see which behaviours result from stimulating these neurons. It pays to be cautious in interpreting the results, especially in mammals, as the techniques may not be as precise as sometimes claimed. Other molecular tools, such as one called Brainbow, 'colour-code' neurons with fluorescent markers so they can be traced through the nervous system.

Hormones and behaviour

Hormones are chemical signals circulated around the body for internal communication and coordination of physiology and behaviour. Hormones share complementary and interacting roles with the nervous system. Hormones may be released in response to neuronal stimulation and hormones in turn can affect neural responses to stimuli. One difference between the nervous and hormone systems is timescale: nerves send messages quickly on millisecond timescales, hormones can have effects over seconds or over longer periods of hours or days. The same or similar molecules can act both as neurotransmitters when released in the synapse (connection) between two nerve cells and as hormones when released into the blood to act around the body. For example, in mammals, *noradrenaline* is a neurotransmitter and *adrenaline* is a hormone. In invertebrates, *octopamine* (the insect counterpart to adrenaline) and *serotonin* act as both neurotransmitters and hormones.

Immediate behaviours and longer term developmental changes can be affected by hormones. For example, when a mother sheep gives birth to her lamb, the hormone oxytocin is released in a particular part of the mother's brain causing her to learn the individual smell of her lamb. In birds and many other animals, hormone release triggered by changing day length stimulates readiness for breeding behaviour at the right time of year.

You can imagine the hormones in the blood as messengers going around a city with the instruction 'ring the bell of any house [cell] with a red or blue doorbell [receptor]'. Hormone messages are only received by cells that have the right receptors expressed on their surface membrane so only they will respond or be influenced. This makes hormones good for multiple effects in different kinds of tissue all around the body. Once it stimulates the cell, the hormone starts a cascade of effects including influencing which genes get activated in the cell's nucleus. A given cell can have multiple receptors in different combinations during its lifetime, so it can be influenced by different combinations of hormones over time. The system can evolve very flexibly in different species by changing patterns of hormone release, by changing receptor specificity, and by determining when particular cells have particular receptors. For example, a key factor in the bonding behaviour of monogamous prairie vole (*Microtus ochrogaster*) males is that they have many receptors for the hormone *vasopressin* on neurons in particular areas of the brain (which are thus activated). In contrast, another vole species, the montane vole (*Microtus montanus*), which is not monogamous has circulating vasopressin but has far fewer vasopressin receptors on those cells in the male brain.

In vertebrates including fish, birds, and mammals, social interactions and other external factors such as day length, together with internal factors such as the animal's fat reserves, form parts of the information integrated by a group of neurons in a crucial part of the brain which controls key hormone releases, the hypothalamus (Figure 7). When the day length is right, for example, special neurosecretory cells in the hypothalamus start to secrete *releasing hormones* to stimulate the pituitary gland to release its own hormones, which in turn activate other hormone-producing glands around the body (such as the *gonads*). Their hormones circulate in the blood and in turn influence the brain, including the hypothalamus, and thus behaviour, by stimulating cells that have the right receptors. All of these

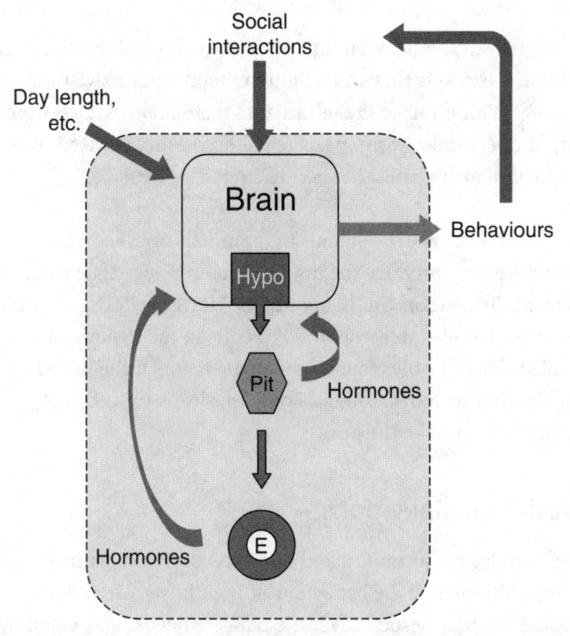

7. In vertebrates, internal and external influences are integrated in the brain's hypothalamus (Hypo), which sends hormone signals stimulating the pituitary gland (Pit) to release its hormones. These activate peripheral hormone-producing (E) tissues, such as the ovaries, testes, and adrenal glands.

actors, from the hypothalamus to the other endocrine glands, are part of interlocking feedback loops that coordinate the animal's behaviour.

Hormonal changes are a key result when a subordinate takes the place of the brightly coloured dominant male in a highly social African cichlid fish (*Astatotilapia burtoni*). Within minutes the newly dominant male starts to show dominant behaviour and skin colours. Neurons in the hypothalamus release a hormone (*GnRH*) which prompts release of reproductive hormones from the pituitary gland. Within hours he is physiologically and behaviourally ready with sperm to fertilize eggs.

In invertebrates, too, social interactions can affect behaviour via hormones. Crickets that win a fight are more aggressive in the next fight. This is due to the release into their blood of the hormone, octopamine. Other hormones such as serotonin released into the blood of losing crickets make them less aggressive.

Hormones influence the signals that animals produce: male house mice only produce the male pheromone signals if there is sufficient testosterone in their systems. Hormone levels also affect responses. For example, the brain circuits in the hypothalamus of male hamsters only respond to female scents if the male is sexually mature and well-fed, as only this leads to sufficient testosterone levels in the brain.

Parasite manipulations

By manipulating hormones and neurotransmitters, parasites can dramatically change the behaviour of their hosts, often to the parasites' benefit—especially if the parasite needs to be eaten by a predator which is the next host in their lifecycle. For example, infection by a parasitic nematode changes the appearance and behaviour of their host, a tropical ant: the abdomens of infected ants turn fruit-berry-red and the ants climb into the tree canopy where they are likely to be eaten by birds, the parasite's next host. Crickets infested with a parasitic hairworm are more attracted to light and are, probably indirectly, more likely to end up jumping into water, where the adult worms emerge to find a mate. Though some scientists question the evidence, it has been suggested that brain infection by *Toxoplasma gondii*, a protozoan parasite, reduces the aversion that rats and mice normally have to cat odour, potentially making them easier to catch. *Toxoplasma* needs to be in a cat to reproduce sexually.

For some parasite–host systems, we are starting to understand the molecular mechanisms that the parasite has evolved to 'hijack' the central nervous system of their host to manipulate the host's behaviour. These findings can be a tool to help understand neural circuits in the host. For example, the parasitic jewel wasp hunts cockroaches (Figure 8). Having found one, the wasp uses selective

8. A jewel wasp (*Ampulex compressa*) female delivers precision doses of neurotoxins as it stings the head of the cockroach (*Periplaneta americana*). The wasp's larva will feed on the living but immobilized host.

neurotoxins to turn the cockroach into a living 'zombie' which it can then guide, like a dog on a lead, to a burrow. The immobilized cockroach serves as a living food supply for the wasp's larva.

The wasp's first sting to the cockroach's thorax temporarily knocks out central motor circuits, blocking motor output to the cockroach's forelegs for one or two minutes so it does not run away. This gives time for the wasp to make a second, longer sting into the cockroach's head. The wasp uses sensors on its stinger to locate a particular part of the cockroach's brain: the command centre responsible for initiating walking. The wasp injects a precise cocktail of neurotoxins directly into it. After the second sting, the cockroach shows a peculiar behavioural change: instead of escaping, it spends thirty minutes grooming itself intensively, most likely because of an insect neurotransmitter (*dopamine*) in the venom. While the cockroach grooms, the wasp goes off to dig her burrow. A further, long-lasting effect of the head-sting is the cockroach's lethargic state, in which, though it is not paralysed, it is unable to self-initiate walking. The head-sting venoms contain *opioids* which target the neurons in the command centre that are specifically responsible for initiating and maintaining walking, without affecting the function of the central pattern generator motor centres. The wasp exploits the separation of central nervous systems into command centres and central pattern generation circuits so she can 'walk' the cockroach, too big for her to move by brute force, to her burrow.

Chapter 3
How behaviour develops

Recipes for success

When we watch birds catching worms and bringing them back to feed their chicks, we can easily overlook the complexity of the behaviours the birds are carrying out so beautifully. How do these and other behaviours develop during the bird's lifetime, from its start as a fertilized egg to being a parent feeding its own young?

The growth of an animal progresses from a single cell (a fertilized egg) to embryo to young animal by the processes of cell division, cell movement, and cell differentiation. These early processes for developing an animal's body and behaviours are the mechanisms of embryology and will continue throughout the animal's life. Using the analogy proposed by Richard Dawkins in *The Blind Watchmaker*, we can think about the orderly process of embryonic development much like the process of baking a cake following a recipe, but where genes provide the recipe instructions. This complicated recipe for embryonic development has millions of steps, with thousands of genes being switched on and off as they interact in coordinated ways. As the steps unfold, different genes are activated (expressed) simultaneously in cells all over the growing embryo. In the right environment, following the recipe contained in the embryo's DNA will lead to the successful development of the animal.

Although each cell is genetically identical to every other cell in the body, the location of the cell and its developmental history determine which genes are activated, and when. For example, cells in the liver express the genes appropriate for liver cells; cells in the heart express heart-appropriate genes. During the lifetime of the organism, individual cells may express different combinations of genes—in response to hormones for example.

Changing a word in the recipe for the cake—for example, a small mutation that changes 'add sugar' to 'add salt'—could lead to widespread changes in how the cake comes out. However, the single changed word is not responsible for the whole cake; rather, the changed word only has its effects in the context of all the other words of the recipe.

Just as *morphology* (or body shape) evolves by natural selection, behaviours evolve in the same way. As genes influence how behaviours develop, selection on behaviour will alter gene frequencies in subsequent generations: genes that lead to successful behaviours in foraging, parental care, or mate choice, for example, will be represented in more individuals in future generations. If conditions change, then mutations of the genes which give rise to advantageous behaviours will be favoured by selection. These mutations can be in the genes themselves (the sections of DNA coding for amino acid sequences to make proteins) or in the enhancer sections associated with each gene which influence when and where the gene will be activated.

Nature and nurture

So far, I have discussed behavioural development in the context of gene activity. However, equally important is the influence of the environment: both genes and environment are intimately and interactively involved. It has been said that trying to allocate aspects of behaviour to only genes ('nature') or environment ('nurture') is a bit like asking whether the area of a rectangle is due more to its length or its width.

The importance of the environment, including the social environment, in behavioural development has long been known from, for example, the way that young animals imprint (learn) their mother's characteristics. What has surprised many of us in recent decades is the discovery that just as genes influence behaviour, behaviour influences gene expression. Using the cake recipe analogy, this means that the behavioural stimuli received by a developing animal and the environment the animal finds itself in can affect the order in which the 'words' (genes) in the recipe are read, as well as whether some words are highlighted for frequent use, silenced temporarily, or silenced for the lifetime of the animal.

Some behaviours do not appear to be learned and have instead been called instinctive or *innate*. For example, a female jewel wasp emerges from its pupa and, though its mother died long before, the wasp is able to successfully complete the complex tasks of locating and stinging a cockroach in precisely the right ways to paralyse it before leading it to a burrow as food for its larva (Figure 8). However, even seemingly innate behaviours have developmental and environmental requirements for full delivery: to say that a behaviour is innate does not mean only genes are involved. This is also true for the development of the senses: a mammal's eyes need to be exposed to visual stimuli during critical periods after birth if the visual cortex in its brain is to develop properly.

Behavioural imprinting

A gaggle of young geese following the ethologist Konrad Lorenz is one of the most famous images in animal behaviour. The young geese had formed an attachment to him because he had ensured that he was the first large moving object they saw during a sensitive period soon after hatching (Figure 9). This phenomenon, called *behavioural imprinting*, is a learning event that occurs in the early development of many mammal and bird species. In the wild, the first moving object a young gosling sees will be its mother, not Konrad Lorenz. The probable function of this filial imprinting is to ensure that the young animal learns to recognize its parents so it can follow them rather than other adults. It leads to a strong social bond between offspring and parent.

The young animal is predisposed to learn during this sensitive period and it is particularly receptive to biologically relevant stimuli including sounds and objects. For domestic chicks and ducklings, the species-specific clucking and quacking sounds made by their mother are highly motivating for following a moving object and imprinting on it. In mallard ducklings, the sensitivity to the mother's quacks depends on having heard earlier, in the egg, the calls made by other embryo ducklings just before hatching. In mammals (and perhaps birds) the individual odours of the parent may also be learned.

In many species there is also sexual imprinting which allows the young to learn how to recognize and choose a member of their species as a mate when they become sexually mature as adults. (When they reached adulthood, Lorenz's male geese courted him!) Depending on the species, these species-specific characteristics can include sounds, visual features, or smells. The importance of

9. A sketch of Konrad Lorenz attracting his imprinted goslings, made by Niko Tinbergen on a visit to Lorenz in Austria.

this learning for adult sexual behaviour varies between species. A nice example is a field study which swopped the eggs of two bird species between nests: great tit (*Parus major*) young become sexually imprinted on the wrong species if reared by blue tit (*Parus caeruleus*) parents. When adult, the fostered great tits tried to mate with blue tits. However, fostered blue tit young were not sensitive in this way: their adult mate choice was not affected by being reared by great tit parents. It might seem strange for great tits to have learning underlie a behaviour as important as choosing a mate of their own species, but in nature great tit young are reared by their parents so normally it works.

Another kind of imprinting is called *familial imprinting* whereby young animals also learn the detailed characteristics of their siblings—including individual appearance, smells, and sounds—to avoid choosing them as sexual partners later in life. This can be demonstrated by cross-fostering animals into another family of the same species. For example, mice avoid mating with familiar siblings, but, if fostered with another family from birth, they avoid their foster-siblings but will mate with genetic siblings they had not met before.

Imprinting can also occur in adults, particularly during parenthood. In the first few hours after giving birth, a mother sheep forms an enduring bond with her lamb. The stretching of the mother sheep's vagina as the lamb is born sends nerve impulses to her brain which set off a cascade of neurobiological and hormonal mechanisms. These create a sensitive period for learning about her lamb. The release of oxytocin prompts maternal behaviour and changes her response to amniotic fluid from avoidance to attraction. She now licks the amniotic fluid off her newly born lamb. While she licks, her brain is prompted to learn her lamb's odours. This learning is important as it allows her to distinguish and suckle only her lamb, not the others in the flock.

Genes and behaviour

While many studies show genetic influences on behaviours such as winter migration routes in birds, it is not easy to identify specific genes with a strong influence on behaviours because most genes are involved in complex webs of interaction with other genes as well as environmental influences. However, some such genes have been found. One was discovered in 1980 when Marla Sokolowski, then a student, was observing *Drosophila melanogaster* fruitfly larvae collected from the wild. She noticed that the foraging larvae could be divided into two kinds: 'rover' larvae that crawled for long distances as they foraged on food (yeast and water paste), and 'sitter' larvae that moved much less. Over the past thirty years, she and others have shown that the difference in behaviour is due to two natural versions (*alleles*) of a gene named *foraging* (or *for*). Larvae with the rover version of the gene (for^R) wander more within and between food patches. Larvae with the sitter version of the gene (for^s) move little and stay within a small food patch. The gene affects adult food-related behaviours in similar ways. As with all other behaviours, while the gene makes a difference, environmental cues are also involved in the behaviour: in the absence of food, rover adults and larvae behave like sitters.

Outside the lab, in a Toronto apple orchard, both *Drosophila melanogaster* variants were found in stable frequencies (70:30 rover to sitter). Why do both versions of the gene persist in wild populations? It seems rovers have an advantage in crowded larval environments, while sitters are selected for in less crowded ones.

What is the *foraging* gene doing to have its effects on behaviour? The gene codes for an enzyme called *PKG* which is expressed in the brain (and other parts of the body). In both larvae and adults there is more PKG expressed in the brains of rovers than sitters. (If PKG levels are raised in sitters, it is this raised PKG level that seems to be responsible for transforming them into rovers.) The *foraging* gene has multiple effects on many other behaviours in larvae and adults, such as learning and memory (rovers have better short-term memory and sitters better longer term memory), as well as other behaviours such as how far from the food the larvae move to pupate. The gene's effects are *pleiotropic*—that is, the gene has multiple effects via interactions with many other genes.

Evolution, co-opted genes, and gene regulation

Once the effects of the *foraging* gene had been identified in *Drosophila*, scientists searched for versions of the gene in other insects. They found it in honeybees and ants, and learnt that the gene was again involved in changes in behaviour related to feeding activity (though in different ways to its action in *Drosophila* larvae). A similar approach led to the discovery that a gene, *period*, which affects the *circadian rhythm* (the daily pattern of activity) in *Drosophila*, has the same role in mammals.

That the same genes can be found in such different organisms, and that many of their functions are also conserved, has been one of the most remarkable discoveries in biology of recent decades. For example, mammals, including humans, have about 20,000 shared genes (99 per cent of mouse genes have a human counterpart). The discoveries came initially from studies on the evolution of development (hence the term *evo-devo*), most famously the highly conserved *homeobox* (or *Hox*) genes which organize the head–tail axis of the embryo in animals as diverse as *Drosophila* and humans. In a similar way, genes that affect behaviour, such as *foraging* and *period*, are used by a wide variety of animals.

But if they share so many genes, what makes animals so different? What causes the big differences in how animals look and behave despite the relatively small genetic differences in their *genomes* (i.e. their total genetic material; their DNA)? The answer lies largely in differences in when and where these genes are *active* (i.e. switched 'on')—changes to the order and timing of reading parts of the recipe, to use the cake analogy (Chapter 2). These customizations of gene activity allow a great diversity of animal forms and behaviours to be generated from very similar sets of genes. Over evolutionary time, genes commonly get co-opted and repurposed by 'teaching them new tricks'.

Switching a gene 'on' or 'off' in a cell is called a regulatory effect. All genes have at least one 'enhancer site', a DNA sequence like a barcode adjacent or close to the gene. Each barcode is a docking site for a specific protein on–off molecule. When the protein docks, this can either switch the gene on (*activate*) or off (*silence*) that gene in a particular tissue. The specific on–off proteins are called *transcription factors* because they control whether the gene is read (*transcription*). Transcription factors are themselves the products of other genes which can be activated or silenced. This means there can be complex webs of fine-tuned interacting signals and feedback loops involving transcription factors, particularly because a gene can have many enhancer sites. Other molecules such as hormones may function by interacting with transcription factors.

You may also see *epigenetic* as a loosely defined umbrella term used in a variety of ways to describe how flexible responses to the environment cause other chemical interactions with DNA said to influence which genes get expressed. The two main proposed epigenetic mechanisms are tagging of the DNA with a chemical flag (*methylation*) and changes to the proteins (*chromatin*) that DNA is wrapped around. These are associated with changes in gene expression, some of which appear to be transmitted between generations. However, many geneticists wonder if the epigenetic changes are symptomatic of differences in gene expression rather than being the cause of changes to gene expression. The main problem is that the epigenetic mechanisms do not seem to have ways for reacting selectively to different DNA sequences to give specificity to which genes are activated. This is in contrast to regulatory elements, which have clear inheritable mechanisms and DNA specificity.

Social influences on brain gene expression

What the animal experiences during its life affects the switching on and off of its genes. These changes allow the animal's behaviour to be fine-tuned to its social and physical environment and the changes allow for considerable plasticity (Figure 10). Complex feedbacks may be involved as the animal's own behaviour will affect its social and physical environment. Behavioural transitions in honeybees and in cichlid fish have been well studied.

In honeybees, social interactions and other stimuli in the nest influence the activation of the *foraging* gene which causes honeybee workers to switch from nursing the brood inside the

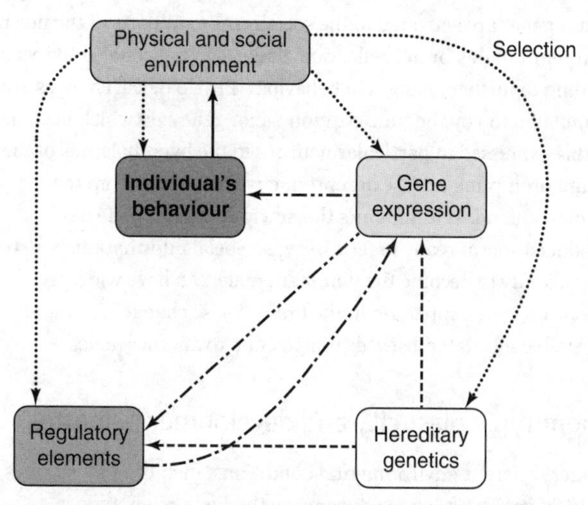

10. The development of an individual's behaviour during its lifetime involves a complex interaction between its genes and its physical and social environment (including other species).

nest to foraging outside for nectar and pollen. The transition to foraging typically occurs at about 2 to 3 weeks of age and involves changes in the expression of hundreds of genes in the brain, as well as changes in hormone levels. *Foraging* is one of these key genes, with increased expression leading to increased production of the enzyme PKG. The timing of the transition from nursing to foraging is sensitive to the needs of the colony, mediated via inhibitory pheromones from older workers. If the older workers are removed, younger workers are induced to switch to foraging earlier (at 1 week old). The precocious workers' brains show increased *foraging* gene expression. Thus the expression of *foraging* and the transition to foraging activities is a response to social information, not the bee's age. The role of the PKG enzyme activity is confirmed by experimentally raising levels in young workers: this causes them to switch to foraging behaviour. The transition of roles is reversible.

Shortly after a subordinate male African cichlid fish, *Astatotilapia burtoni*, wins a fight to become the top male, many transcription factor genes are activated in the socially relevant parts of the newly dominant male's brain, reflecting changes in neuronal activity and leading to further changes in behaviour. Physiological changes are stimulated too by the transcription factor gene *egr1* which becomes highly expressed in particular neurons in the hypothalamus of the brain, prompting release of reproductive hormones from the pituitary gland. Within hours the newly dominant male has produced sperm ready to fertilize eggs. Social information, here the opportunity to become the dominant male, can have widespread effects on gene expression in the brain. These changes are rapidly reversible if he later loses his top position to another male.

Phenotypic plasticity and developmental paths

Under different environmental conditions, including social ones, a single genotype can produce more than one phenotype in the form of different morphologies and sets of behaviours. The reversible changes in the male cichlid fish are one example.

In honeybees, remarkable and irreversible changes happen in the development of female larvae, depending on what they are fed. Each female egg has the potential to become either a long-lived fertile queen or a short-lived sterile worker. The environmental switch is a protein-rich 'royal jelly' produced by nurse worker bees. If a female larva is exclusively fed this royal jelly, genes will be switched on to make her a queen. If fed mostly pollen, other patterns of genes will be switched on and she will develop into a worker honeybee. By the time she emerges as an adult after pupating, she will be a worker or a queen with the expected characteristic behaviours and activity. There are big differences between workers' and queens' brains, neural circuits, and resulting behaviours.

The desert locust, *Schistocerca gregaria*, can famously change between its solitary and gregarious phases (forms) in response to population density. When gregarious, it forms vast swarms of millions of locusts covering thousands of square miles. At low densities, in its solitary form, it avoids other locusts and looks so different from the gregarious form that scientists thought the two forms were different species. The switch from solitary phase nymphs into gregarious phase nymphs can be triggered by many factors, including the combination of sight and smell of other locusts, or being repeatedly touched on their hind legs. These stimuli cause the release of the hormone serotonin in particular parts of the central nervous system, which initiates the behavioural changes, followed by changes in gene expression throughout the nervous system and the rest of the body.

More subtle effects of the environment and upbringing on the behaviour of animals can also occur. This was demonstrated nicely in a study observing variation in the adult behaviour of genetically identical mice. Inspired by human twin studies, researchers took forty genetically identical lab mice, all adult females, and placed them together in a large enclosure with multiple levels and lots of places to explore. An electronic tag in each mouse allowed its movements to be tracked over three months. The study measured how much the mice explored and used the environment (a good measure of their social activity): it found enormous variation between mice and that these differences increased over time.

Even at the start of the experiment when the mice were first released into the enclosure, the mice were already potentially different from each other, despite being genetically identical. These differences could be the result of a variety of early-life environmental factors, all with potential regulatory effects. For example, because the mice came from many different mothers, they could exhibit different maternal effects from their mother's nutrition, parenting ability, and social status. Their individual position in the uterus, as well as the sex of adjacent embryos, could affect their exposure to hormones during development. Once born, early experiences such as competition for suckling, and winning or losing fights in the nest, could offer different experiences within litters, quite apart from the differences in maternal care between litters. It is surprising how differently these genetically identical animals developed in the same environment. The experiment is a reminder of the importance of social and other environmental effects on the development of behaviour.

Sex determination, and behaviour of males and females

While the behaviours of males and females in most species are different though overlapping, there is an enormous diversity of sex determination mechanisms. In humans and most other mammals, sex is determined by sex chromosomes (XX females; XY males). However, male and female embryos start with the potential for either masculine or feminine behaviour. Which behaviour develops depends on whether a particular hormone pulse occurs during late embryonic development. The *Sry* gene found on the Y-chromosome of male mammals leads to the development of testes rather than ovaries in the embryo. During a brief, critical period the testes secrete a short pulse of testosterone. This *organizing* pulse switches neurons in the hypothalamus towards masculine behaviours for life. Female's brains are not masculinized in this way because they do not have testes to produce the testosterone pulse. The hypothalamus is a key region of the brain involved in regulating behaviour and hormone release (Figure 7). When adult, day to day levels of testosterone affect the male's behaviour.

Unlike mammals, a surprising proportion of other vertebrates rely on environmental or social cues for developing into male or female. Among these are temperature-dependent sex-determination in many reptiles including alligators, crocodiles, and some turtles, with females produced at high and low temperatures, and males at intermediate temperatures. Many fish species change sex during adulthood. For example, the coral reef anemonefish (*Amphiprion*) changes from male to female. Anemonefish live in groups in a size-based hierarchy. The largest individual is female and her male mate is the second largest. If she dies, the male changes sex to become a female; and the next largest male, previously non-breeding, becomes the new male. (In the animated cartoon *Finding Nemo*, Nemo loses his mother and his father searches for him. In reality, Nemo's father would have stayed home and changed sex to become female.)

In the fruitfly *Drosophila*, sex determination works differently from mammals. While *Drosophila* has XY/XX sex determination, sex is not determined by hormone pulses. Instead, sex determination genes act during development at the level of the individual cells in which they reside, making it possible to artificially create a mosaic animal with both male and female cells. The effect on the fly's behaviour of having particular neurons male or female can be investigated. The key genes responsible for the development of the different male and female neural circuits underlying sexual behaviours are *doublesex* and *fruitless*, acting together. Both genes produce transcription factors, in male and female versions. *Fruitless* is found only in insects but a family of *doublesex*-related *Dmrt* genes occurs throughout the animal kingdom, playing a fundamental role in sex-specific differentiation in all animals studied so far (for example, responding to the gene *Sry* in mammalian sex development).

Effects between and across generations

Non-genetic, parental (intergenerational) effects on offspring occur in many species. For example, gregarious-phase female desert locusts secrete a *gregarizing factor* in the foam around their eggs which makes their nymphs more likely to become gregarized themselves. In mammals, developing embryos can be affected by their mother's diet, physiology, and social status. This can lead to different patterns of gene activation in embryos, potentially with lasting effects on the physiology and behaviour of the young. Similarly, differences in mothering received in one generation can affect the next. For example, the amount of licking and touching young mice receive, with lasting effects on their gene expression, may affect their own parental behaviour when adult, with corresponding effects on their offspring.

Bird song as a model system

How songbirds learn their songs brings together many of the themes of this chapter. There are many similarities between the ways that songbirds learn their songs and human babies learn language, namely through a combination of predispositions and learning. Young songbirds, such as zebra finches and canaries, learn their song during a sensitive period by imitating the songs of nearby older adults of their own species. They ignore the songs of other species. After memorizing the song of adults, the young bird starts by producing a highly variable *subsong* which has been likened to human infant babbling. This subsong is refined by comparing it with their internally memorized *template* and modifying their own song until it matches (Figure 11). Babies go

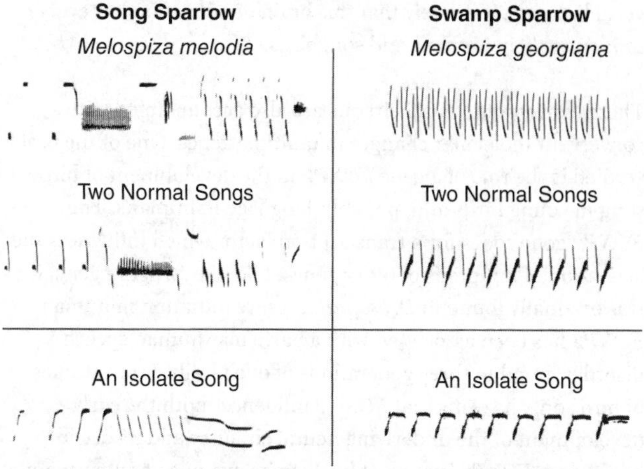

Song Sparrow
Melospiza melodia

Swamp Sparrow
Melospiza georgiana

Two Normal Songs

Two Normal Songs

An Isolate Song

An Isolate Song

11. Sparrows which can hear their own species' songs during development produce normal songs themselves as adults, visualized in sonographs (top). Isolated birds which do not hear their species' song during their sensitive period produce abnormal songs (bottom).

through a similar process of listening before babbling and then gradually perfecting their speech. Like songbirds, humans have a sensitive period early in life when language learning is most effective. Until puberty, and particularly when younger, humans seem to have a remarkable predisposition to learn whichever of the 6,000 or so human languages they are exposed to, without formal teaching.

There are similarities too in the organization of the interconnected brain regions involved in learning and producing song in birds and language in humans. Songbirds have separate specialized brain regions for sound recognition, memory, and producing song. In human infants there are suggestions that interacting brain regions for perception, memory, and production of speech may be organized similarly. As mammals (including non-human primates) and most birds do not show these kinds of complex vocal learning, it is likely that this brain organization has evolved *convergently* in humans and songbirds.

These convergent neural circuits are also accompanied with convergent molecular changes in multiple genes. One of the best studied is the role of a gene *FOXP2* in the development of bird song in young birds and, possibly, language in humans. The *FOXP2* gene codes for a transcription factor which influences the activation of hundreds of other genes. Like the *foraging* gene, it was originally found in *Drosophila*. A rare mutation in human *FOXP2* has been associated with a particular human speech disorder found in three generations of one family. From studies of bird song, it seems that *FOXP2* influences both the early development of the underlying neural circuitry and its use in adulthood. *FOXP2* is at most likely to be just one of many, many genes involved in developing our brains to learn and speak human language. Studies of gene influences on bird song may nonetheless help our understanding of human language.

A further discovery leading from studies of bird song is that adult brains in vertebrates are much more plastic than previously thought. Studies of adult male canaries revealed that brain regions responsible for song grow larger in the spring due mainly to the development of new neurons from neural stem cells, a process called *adult neurogenesis*. This is associated with the development of a new song repertoire each year. In mammals, adult neurogenesis is probably confined to just two regions—the *hippocampus* and a region that supplies new neurons for the olfactory system (which is constantly renewed throughout life).

Play

I end this chapter with another good example of behavioural development—play—though we still do not understand its functions, which may vary from species to species.

When kittens play with a ball of string or rough-and-tumble with each other, we recognize it as play behaviour, similar to that of human children. Play seems distinct from other behaviours, and it seems to be done spontaneously for its own sake, repetitively, and often appears exaggerated. Play seems to be a common feature of the young of most mammals and a few birds such as crows. (Some researchers have suggested that play occurs more widely across the animal kingdom.) Many mammals have a *play solicitation behaviour*—in young primates (and humans) there is a relaxed open-mouth 'play-face' which signals that the individual's next actions are play, not for 'real'. Adults may play too. Dogs do a bow-down display to each other before they start play-chasing or play-fighting; dogs make the same crouching bow-down display to encourage you to throw a stick again. Adult bonobo chimpanzees use the play-face in playful interactions with other adults and with juveniles. Play has its own rules so that, for example, play-partners take turns to chase each other.

Play can be costly because it takes energy and time which could be spent foraging. While distracted in play, the young animal may be at great risk of being caught by a predator. For example, 86 per cent of young Southern fur seals eaten by sea lions were play-swimming with others when they were caught. Against these costs many functions have been proposed for play, including practice for adult behaviours such as hunting or fighting, and for developing motor and social interaction skills. While these and other theories seem plausible, there is little experimental evidence in any animal. For example, detailed studies which tracked juvenile play and adult behaviour of meerkats showed no evidence that play-fighting influenced fighting ability as an adult. The persistence of play across so many animal species, despite its costs, remains a mystery. The answers are likely to involve diverse and multiple factors, which may be quite different in different species, as might what we call *play* itself.

Chapter 4
Learning and animal culture

All animals learn

Learning is an animal's capacity to change behaviour as the result of individual experience, so that their behaviour is better adapted to its physical and social environments. Learning helps to fine-tune behaviour in flexible ways that would be difficult to encode genetically. It is an inherent property of nervous systems in even the simplest animals, including the sea slugs used as models for neural circuits (Chapter 2). Memory is an animal's capacity to retain learned information to influence future behaviour.

The ideas of classical *conditioning*, one kind of learning, were made famous by Ivan Pavlov in the 1920s, with his studies of dogs' behaviours when a bell was rung before feeding. After the dogs had learned the association of a bell ring being followed by food, the dogs would salivate to the sound of the bell alone, without food. Another kind of learning is *trial and error*, in which the reward is linked to an action that the animal itself has made by chance. If a certain behaviour leads to a reward, it will be repeated whereas behaviours which are not rewarded will occur less and less. An experimenter can exploit this by giving a reward each time an animal behaves in the way the experimenter wishes to encourage. (This kind of trial and error learning is also called *operant conditioning*. Operant conditioning is the basis of the *Skinner Box* in which a pigeon pecks a screen to get a reward.)

In the middle of the 20th century, these theories of learning stimulated what has been described as the heroic period of American experimental psychology. This period was characterized by ever more detailed and sophisticated studies of learning in lab rats and pigeons, using apparatus like mazes and levers.

The ethological approach to animal behaviour has shifted the focus to the behaviours of animals in the wild, and how learning underlies the ways animals recognize their young, find food more efficiently, avoid predators and poisons, and countless other tasks related to their ways of making a living in the natural world. Many studies have further investigated these behaviours in laboratory settings, but with ecologically relevant questions at their heart.

Learning gives flexibility to behavioural responses

What members of a species learn is an evolved trait, strongly related to their ecology and life history. For example, generalist-feeding animals show *one-trial learning*. Such animals can feed on many different kinds of foods and need to be flexible enough to try all the possible foods that might be good to eat, while avoiding eating poisonous food more than once. For example, adult rats eat only a small quantity of any unfamiliar food. If they become ill in the hours after a meal, rats learn, after a single exposure, to associate the smell of the offending food with the negative outcome and will reject any food with this odour in future. This *learned taste aversion* is common across the animal kingdom, including snails, insects, and humans (for years after a teenage excess, I could not face the smell of gin). Interestingly, while rats have an evolved ability for a learned taste aversion prompted by later illness, they do not easily learn to associate tastes and electric shocks. This is an example of a species-specific constraint on learning and also shows the importance of asking ecologically relevant questions in experiments.

Birds learn to associate the warning colours of poisonous butterflies with later ill effects, avoiding these butterflies in future. This learned avoidance inadvertently strengthens selection for the warning signal in target species and it also selects for potential mimic species to produce these colours. Generalist-feeding bat species readily learn taste aversions. However, vampire bats, which feed only on vertebrate blood (a non-toxic food), do not learn taste aversions, supporting the suggestion that learned taste aversions are an evolved ability of generalist feeders.

A readiness to learn may be the default mode for animals. However, innate responses that do not require learning may evolve in some situations. For example, a danger that remains the same from generation to generation, and where one mistake might be fatal, provides little opportunity for learning. In the American tropics, brightly coloured poisonous coral snakes (*Micrurus surinamensis*) present a lethal danger to kiskadee birds (*Pitangus sulphuratus*). Young hand-reared 'naïve' kiskadee birds, with no experience of coral snakes, nonetheless give alarm calls and avoid models painted with brightly coloured red, yellow, and black bands that match the colours of coral snakes. Naïve young blue jays and house sparrows, species from temperate habitats where the snakes are absent, show no such avoidance or alarm.

Hiding food for later

Some birds show astonishing abilities for learning and memory in the way that they retrieve seeds they have previously hidden as a food bank.

During the late summer and autumn, Clark's nutcracker (*Nucifraga columbiana*), a member of the crow (corvid) family, spends every daylight hour collecting and storing whitebark pine seeds. It collects up to 150 seeds in a special pouch under its tongue before it flies off to store them (Figure 12). A single bird will hide more than 30,000 seeds in groups of three to four seeds in 2,500–3,000 different locations over an area of up to 100 square miles. The bird relies on its stored seeds to survive the tough winters in the alpine conifer forests of western North America. The behaviour is called *scatter-hoarding*, as the seeds are hidden in many scattered sites rather than in a concentrated 'larder', which would be more vulnerable to theft if discovered. Because not all seeds are retrieved, the trees benefit: the whitebark pine depends wholly on Clark's nutcracker for dispersing its seeds. European jays similarly disperse the acorns of the English oak.

Scatter-hoarding birds remember the location of each stored item, and can also remember if they have already retrieved those seeds.

12. A Clark's nutcracker hiding pine seeds in a food cache. Its crop is filled with up to 150 seeds collected from whitebark pine cones as far as 20 miles from the cache sites.

Clark's nutcrackers can remember where they have hidden food up to nine to ten months later. Laboratory experiments have shown how they use landmarks to relocate the hidden seeds. In an experiment, a Clark's nutcracker was allowed to hide seeds in a sandy arena that had logs and rocks at each end as landmarks. Before the bird was allowed to return to the arena to retrieve seeds, the landmarks at one end were moved 20 cm further out. At the 'moved end', the bird was misled by the moved landmark into consistently searching in locations that were offset by about 20 cm. However, at the end with the untouched landmarks, it went to the correct locations. Nutcrackers seem to use bearings from multiple landmarks, the more the better, allowing them to relocate stored seeds even under the snow.

The part of the bird and mammal brain important for memory, especially spatial memory, is the hippocampus. As you might predict, scatter-hoarding bird species have a proportionately larger hippocampus than related species that do not show this behaviour. The size of the hippocampus in scatter-hoarding birds changes with the seasons, getting larger as hoarding starts, either in anticipation of or in response to the greater brain activity required to memorize the storage locations.

Some humans take on especially challenging spatial memory tasks too. For example, trainee London taxi drivers must memorize 25,000 roads, as well as the locations of thousands of places of interest. You can often see trainees on mopeds, acquiring 'The Knowledge' needed to pass the exam. Brain scans at the end of three to four years of training showed that the size of the hippocampus had increased in the brains of trainees who passed the exam, but not in the brains of men used as control subjects or in trainees who did not qualify. (Brain scans had shown no differences between the three groups at the start of the observations.) As might be expected, on tests that measured memory of the distances between London landmarks, the successful trainees performed better than the controls or the trainees who did not qualify. However, the successful trainees were poorer on remembering a complex drawing, which suggests that their increased spatial knowledge or processing might come at the expense of other memory capabilities.

Learning bees

A honeybee has a brain that is smaller than a pinhead, but size is no barrier to producing sophisticated and complex behaviours, including many kinds of learning and memory, comparable to many vertebrates.

Learning plays a large part in making honeybees such efficient foragers. When honeybees find a profitable flower patch, they learn the location so they can find it again. They also learn the location from the dances of recently returned foragers (see Figure 17 in Chapter 5). Because some flower species are better sources of nectar, honeybees learn the shape, colour, and scent of the flowers that are currently rewarding. (They even learn the best time of day to visit each flower species, as well as the location of its patch.) Learning is rapid and accurate: in the first four to five seconds of a nectar sip, the bee can learn to associate the nectar reward with the features of the flower. This association is remembered for days, and, if the learning is repeated, the association will influence the bee's choices for the rest of its three-week-long life of foraging.

A hungry bee will continue to learn if it is put in a 'harness' that allows free movement of the bee's antennae and *proboscis* (tongue tube) (Figure 13). When a sugar solution is touched to the bee's antenna, the bee will reflexively extend its proboscis. If a sugar solution is touched on the tongue as a reward, the bee will associate the reward with whatever odour is being blown over its antennae at the time. This is classical Pavlovian conditioning, as the odour alone will now cause the bee to put out its tongue even if there is no sugar reward.

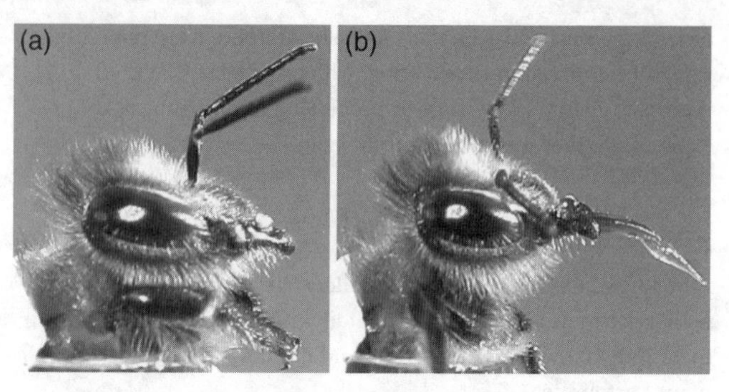

13. Honeybee in a harness showing proboscis retracted (a) and extended (b) in response to an odour it has been trained to associate with a sugar water reward.

By studying the learned proboscis extension response, researchers can explore the different stages of memory formation, as well as the places in the bee brain where memories are stored as a *memory engram* (the representation of stored knowledge which can be used for mental operations and behavioural control). The engram is stored as lasting changes in the strength of the connections (the synapses) between specific neurons which integrate sensory inputs and motor outputs. In the honeybee brain, these neurons are located in the part of the brain called the *mushroom body*, which is the insect equivalent of the mammalian hippocampus and *prefrontal cortex*. Engrams have also been studied in birds using the behavioural imprinting first explored by Konrad Lorenz (Chapter 3) as a tool. The visual imprinting response of the domestic chick has allowed researchers to locate memory traces to a particular part of the bird *forebrain*, providing insights into memory formation in vertebrates.

Social learning and culture

Learning from other individuals is called *social learning*. Across generations this leads to *cultural traditions*.

Cultural traditions explain the development of local song dialects in songbirds as the young birds imitate the songs of nearby older adults of their own species. Intergenerational cultural transmission also explains why successive generations of bluehead wrasse (*Thalassoma bifasciatum*), a coral reef fish, spawn daily in the same place, even though many other sites appear to be just as good. This was demonstrated by removing all the fish and replacing them with bluehead wrasse from elsewhere on the reef: the introduced fish ignored the old site and chose their own site for spawning—which then became the new traditional site.

One of the most famous examples of cultural transmission is the invention of washing of sandy food by a young female Japanese macaque (*Macaca fuscata*) in the wild. Japanese researchers had placed the sweet potatoes on a sandy beach during a long-term study of the behaviour of the monkeys. Key to the observations was that the researchers were following the behaviour of the monkeys as individuals, each of which had a nickname. It is significant that it was a young monkey that first took a sandy sweet potato to the water's edge and washed off the sand. The behaviour was copied by other individuals, especially younger juvenile monkeys. Initially, some mothers learnt from their offspring. Within a few years almost all the monkeys in the troop washed their food in this way, having learnt as infants.

Another well-known example of animal innovation and culture occurred in the 1950s when British birds, including great tits, began piercing the foil caps of milk bottles to take the cream (which in those days collected at the top of the bottle). Again this was a behaviour which spread by imitation.

Is it possible to artificially create a tradition in the wild? Yes. Scientists working in an Oxford oak wood taught a few male great tits one of two different—but equally effective—methods of opening a puzzle-box to obtain food. Afterwards, each male bird was released back into a different part of the wood, along with three automated puzzle-boxes that could be opened by either method. Would birds in their part of the wood learn the opening method from their local, released 'demonstrator' male? All the wild great tits at the release sites had individual electronic tags which allowed the researchers to know which method each bird had used to open the puzzle-box. Within twenty days, 75 per cent of the wild birds had learnt to open the puzzle-box—but the method they used was the one 'their' local demonstrator male had been taught. The newly introduced local traditions persisted over at least two generations, with young birds likely to learn the method they saw used by their local adults. Wild birds at control sites with a male which had not been taught any method showed little success.

While social learning is common, teaching in a human way does not seem to occur in non-human animals. One exception may be the way that adult meerkats apparently teach pups how to handle live prey—in particular, venomous scorpions. The adults first bring the pups dead scorpions, then, as the pups get older, they bring live scorpions with the sting removed. Finally, they bring the pups intact, live scorpions. The adult monitors the pup's handling behaviour, and will nudge prey and retrieve escaped prey. Pups learn the skills more quickly if presented with live scorpions rather than dead ones. How do the adults match the prey to the pup's age? Playback of recordings of calls made by different aged pups showed that adults seemed to just match the prey to the begging calls the pups make, which change as they get older. Whether the adult meerkats are really 'teaching' is questioned by some researchers, as the adults do not demonstrate techniques to the pups. In addition, the adults have not been shown to respond to an understanding of what the pups can and cannot do, which is a normal expectation in human teaching.

Cumulative culture, the way that cultural traits 'ratchet up' in complexity over time, may be a distinctive human phenomenon. Cultural transmission in chimpanzees has not allowed them to develop more complex tools than the ones they have already used for tens of thousands of years. By contrast, human culture allows us to develop and build tools into things much more complex than could be invented by any one individual. In a search for possible mechanisms of human cultural transmission that might explain some of these differences, lab experiments have compared the opening of multistep puzzle-boxes by small groups of chimpanzees, capuchin monkeys, and young children. Groups of children were much more successful in progressing sequentially through the more difficult puzzle steps than the other primates. Key factors in the greater success of the children included greater cooperation and imitation, as well as human language for explaining ideas. All these may be important for the phenomenon of cumulative culture.

Much of the interest in whether animals can learn from each other, whether they have cultural traditions and use tools, or whether they use reasoning and language, goes back to fundamental questions about which qualities and behaviours are characteristically *human*, and what, if anything, separates us from other animals, including non-human primates. Growing evidence from studies of animal behaviour shows that many of our abilities are matched by other animals.

Using tools

We once thought that using tools was a uniquely human ability. However, a wide range of animals has now been observed using tools—for example, hunting wasps use a stone held in their mandibles to compress the soil as they close a burrow; and Galapagos woodpecker finches use a cactus spine to extract insects from under tree bark. Also included in some lists of non-human tool users are archer fish that jet water to bring down insect prey from perches above water.

Some animals are able to use a variety of different tools. Common chimpanzees (*Pan troglodytes*) break open hard nuts with stones, strip leaves from sticks to form a termite fishing tool, and use strong sticks to dig for edible roots or underground bee nests. They also tear and chew leaves to make a leaf-sponge to drink water from tree holes. Because different chimpanzee populations differ in their tool use, these examples are likely to be cultural traditions. Other primates also use stone tools: in Brazil, capuchin monkeys use stones as tools to crack nuts and to dig; and in Thailand, long-tailed macaques use stones to open oysters and other seashore prey.

Wild New Caledonian crows (*Corvus moneduloides*) use many kinds of tools to extract beetle grubs from rotten logs (Figure 14). They fashion one kind of tool from twigs by removing a side branch to leave a hook. Another kind is made by tearing a strip from a particular leaf that has a row of thorns on its edge. Hand-reared New Caledonian crows also show a propensity for tool use. Other corvid bird species show an intriguing ability to use tools in laboratory experiments but do not use them in the wild, perhaps because using tools in the wild provides no advantage for

14. A New Caledonian crow uses a tool fashioned from a stick to extract a beetle grub.

normal food gathering. For example, in the laboratory, rooks (another kind of crow), which do not usually use tools, dropped stones into a narrow tube of water to raise the water level enough to allow the bird to reach an insect floating on the surface.

What are they thinking?

The widely available videos of chimpanzees and New Caledonian crows using tools to gain access to food are remarkable and compelling. It is tempting to ascribe insight to some of their responses. However, labelling a behaviour as 'insight', with the implication of an 'aha moment', does not help us understand how the task is achieved, which might be the result of simpler thought processes. In some cases previous experience matters. For example, the rooks dropping stones had previously been trained to nudge stones into a tube (or had seen other rooks doing it) in a different context.

While humans tend to rate *flexible* intelligence and problem-solving highly, *specialized* intelligence—such as learning and retracing a migration route—can also result in effective problem-solving of a particular kind. For example, Clark's nutcrackers are exceptionally good at the specific task of relocating hidden seeds. However, on other tasks, such as colour-based learning tests, nutcrackers are no better than other species in the crow family.

More than thirty years of patient work with African grey parrots (*Psittacus erithacus*) have demonstrated that these birds can handle abstract concepts, and acquire and use English speech in ways similar to very young children. Irene Pepperberg devised a very effective way to teach, via social interactions, one particular parrot, Alex, to associate words with objects and their properties such as colour. Alex could request the objects by name and could answer simple questions in English about them. For example, Pepperberg could show him a particular object and ask him: 'What colour?' (his reply, green), 'What shape?' (four-corner),

'What matter?' (wood), and 'What toy?' (block). To respond in this way Alex needed to have an understanding of these labels and the concepts of colour, shape, material (matter), and toy. He could also work with more abstract concepts such as bigger/smaller, same/different, and absence. He could correctly count up to six objects and provide the number name to represent the quantity, and he spontaneously produced the idea of 'none'. The videos of his interactions with Pepperberg are fascinating.

Honeybees also have remarkable abilities to learn abstract concepts, such as the difference between symmetry and asymmetry; above and below. Honeybees can also learn the concepts of sameness and differentness. As they fly through a maze, the bee can learn to turn towards a symbol that was the same as (or different from) the symbol at the entrance to the maze. A bee can also learn to transfer a concept to other stimuli, so, for example, if it learns a concept regarding two colours, it can apply the same conceptual rule to a choice of two patterns. Interestingly, with bees we are less tempted to invoke human-like intelligence in the ways that we would with a bird or mammal.

Some of the challenges to understanding animal minds come from our anthropomorphic inclinations to assume that an animal is thinking as we might in that situation (consciousness in animals is discussed in Chapter 8). For example, it is easy for us to imagine that some non-human animals have a *theory of mind* similar to ours, where an individual can understand that another individual has its own intentions or desires. There have been many detailed experiments to explore this possibility, notably in chimpanzees and corvid birds such as western scrub jays (*Aphelocoma californica*). For example, Nathan Emery and Nicola Clayton looked at scrub jays hiding food to eat later. Unlike Clark's nutcracker, scrub jays are social and there is a risk that another individual will observe the bird hiding food, note the location, and steal it later. In the laboratory experiments, a jay could hide food in private or in full view of another jay. When the jay had been observed by a potential thief while food-hiding, the jay would be more likely to re-hide the food in new places. However, naïve birds which had not themselves been thieves did not re-hide their food when they were observed hiding it. These results suggested that the jays with experience of stealing food could relate their experience to the possibility that their own hidden food was at risk of theft if another bird saw the hiding taking place.

These kinds of apparent *mind-reading* and other observations might suggest human-like insight and intelligence. However, Sarah Shuttleworth reminds us that the challenge is to distinguish reasoning about another's mind from responses to behavioural cues alone. Simpler mental processes such as associative learning, responses to behavioural cues, and species-typical predispositions may account for many of these complex behaviours, even though such explanations might seem rather 'killjoy' by comparison.

Ironically, our own behaviour is often poorly accounted for by anthropomorphic explanations. We often use rules of thumb and other simpler, unconscious, and sometimes irrational, processes that we share with other animals. It could be that valuing and studying these simpler mechanisms in animals might help us to better understand ourselves.

Chapter 5
Signals for survival

Getting the message across

Animals communicate in myriad ways. Male birds sing to attract mates and repel rivals. Brightly coloured poison-arrow frogs advertise their toxic protection to potential predators. Male fiddler crabs wave their claws to court females. Usually, when animals communicate, the receiver of the communication responds by changing its behaviour. In other cases, which are harder to study, the response of the receiver may involve slower hormonal changes affecting physiology and behaviour. For example, female ring doves (*Streptopelia risoria*) that are kept together do not ovulate. However, just seeing a male dove's courtship display, even behind a window, is enough to activate a part of the female dove's brain, the hypothalamus, which stimulates her pituitary gland to release hormones (Figure 7). These hormones cause her ovaries to secrete oestrogen, which, in turn, stimulates ovulation and promotes nest-building behaviour.

Communication can use any of the senses including vision, hearing, smell and taste, touch, and electric senses. Signals can also involve more than one sense: some jumping spiders' male courtship signals combine *visual semaphore* (waving their forelegs) and complex seismic vibrations sent via the leaf they share with the female.

Signals have evolved to alter the behaviour of another organism (the receiver), and signals work because the receiver's response has also evolved. Animals also respond to cues; for example, mosquitoes are attracted to the cues of human smells and body heat. Cues can be distinguished from signals because, as in this example, although the mosquitoes have evolved highly sensitive receptors for detecting our smell and heat, our smells and body temperature have not evolved as signals to attract mosquitoes. We give off smells and heat just by being alive—we would rather that mosquitoes *didn't* find and bite us.

Signals can evolve from cues if responding to such cues gives a selective advantage. For example, goldfish sex pheromones are related to female goldfish hormones. So how did hormones become adopted, co-opted, as pheromones? A likely explanation is that, way back in evolutionary time, as females developed eggs for release at spawning, hormones leaked out through their gills and in their urine. Any mutant males that could better smell these molecules (smell works just as well under water as in air) would have a selective advantage because they would get to the female first when she released her eggs. Males were selected for greater and greater smell sensitivity for these molecules, as well as for greater receptor specificity to avoid being stimulated by similar molecules that might give false alerts. Then gradually, over evolutionary time, females that produced more of the molecules—which would now be a signal—would be selected for too, as a male would be more likely to be attracted to fertilize her eggs.

Another path for the evolution of signals is by exploiting a pre-existing sensitivity in the receiver. Tropical anole lizards are sit-and-wait predators which watch for moving insects. Their brains are tuned to visually detect walking insects by their rapid jerky movement, in contrast to the gently moving leaves in the background. To attract females and defend its territory, the male anole displays by rapidly bobbing its head and brightly coloured throat-fan up and down. The display starts with head-bobs at just the right speed to stimulate the visual-motion detection circuits in the brain of the receiver. The signal display seems to have evolved to exploit the receiver sensitivity to prey movement, thus getting the attention of the receiver.

Signals evolve so they transmit effectively to the receiver, given the physical constraints of the signaller and habitat. Limited by the laws of physics, plant-sucking bugs smaller than 1 cm can only produce high frequency ultrasonic sounds in air, which would quickly attenuate. Instead, like the jumping spider, they send very effective vibratory courtship messages along plant stems. For visual signals, contrast with the background in the ambient light in that habitat as well as the colour wavelength sensitivities of the receiver's eye become important factors for signal evolution. This has been explored in four species of manakin (*Pipra* species), black feathered birds which make courtship displays at different heights in the Amazonian rainforest of Venezuela. Males of the two species that display in the gloom of the forest floor have blue and red head crests which contrast well with their background, whereas the mid-height golden-headed manakin ensures an eye-catching display by moving into a light gap. The species highest in the canopy has a white crest to maximize contrast in bright sunshine.

Honest signals

Across the animal kingdom, signals within a species seem to be honest on the whole. This might seem surprising: what is to stop animals from boasting or deceiving when they communicate? One reason is that, on average, signals must communicate useful or truthful information to ensure that the receiver does not evolve to ignore the signal—like the villagers in Aesop's story of the boy who cried 'wolf'.

A variety of additional mechanisms seem to ensure that most signals are honest. The first mechanism applies to signals which reflect an inherent feature (*index*) of the signaller, which cannot be faked. In the spring, European common toads (*Bufo bufo*) gather in ponds to spawn. Males pair with females by gripping their backs. Unpaired males try to dislodge the males in such pairs so they can take their place. When a paired male is challenged by another male attempting to push him off the female, the paired male croaks. His croak gives a reliable signal of his size: larger males have longer vocal chords that produce a deeper pitched croak. To test this, researchers temporarily silenced some paired males by placing a rubber band over their mouths. Then, while a medium sized male attempted to dislodge him from the female, the researchers played back the sound of either a deep or high pitched croak next to the silenced male (Figure 15). Whether the size of the silenced paired male was large or small, fewer

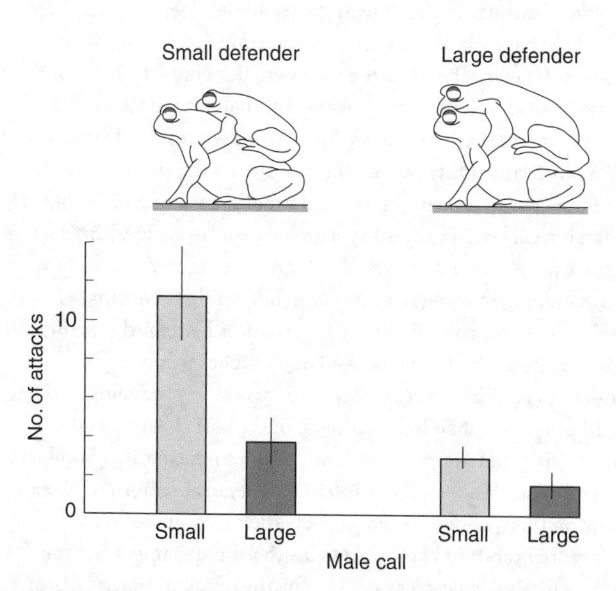

15. Male toads are more likely to attempt to dislodge a male, large or small, on top of a female if they hear a recording of the croak of a small male.

16. Plumage variability in Harris's sparrow. The darker males are dominant in the flocks and win most of the fights.

attempts were made if the deep croak was played. It seems that the rival males are using the croak as a signal to assess the likelihood of success: a large male with a deep croak will be harder to dislodge.

Another mechanism that keeps signals honest is the potential cost of cheating. In the breeding season, red deer stags (*Cervus elaphus*) have roaring contests to determine ownership of a harem. It seems that, like the toad's croaks, characteristics of the stags' sounds are reliable indicators of their size. Because the smaller animal usually backs down, these contests only rarely lead to a physical fight. However, there is always the risk that a fight might occur that could injure an animal pretending to be bigger than it really is. Many bird species have 'status badges'. In Harris's sparrows (*Zonotrichia querula*), high-ranking individuals have dark patches of throat feathers as a 'bib', while subordinate individuals are pale (Figure 16). If a subordinate bird is artificially given a dark bib, it is challenged and fought by dominant individuals. While there is little energy cost to producing dark feathers rather than light feathers, the cost of cheating is the aggression that illegitimate claims of dominance bring. Similarly, paper wasps (*Polistes dominulus*) have a facial pattern of black dots on their 'upper lip' that reflects their dominance status. If experimenters add black dots to a subordinate's upper lip, the 'imposter' is subject to aggression by the dominant members of the group. In both Harris's sparrows and paper wasps, it appears that the animals use the colour badge to assess individuals' status.

Cheating is pointless when the benefits from a signal are mutual, when there is a common interest. Species-recognition signals which allow males and females to recognize each other as members of the same species do not offer any advantage to deception. It is in the interest of both sides to be honest, as mating with a member of another species would be pointless, with no fertile offspring produced. Within a honeybee hive, the bee dance to indicate the direction and distance of good nectar sources is likely to be honest as all the bees are part of the same colony, with a shared interest.

Honeybee dances

Honeybees (*Apis mellifera*) use a *waggle dance* to communicate the location of rich flower patches to their nestmates. This *dance language*, decoded by the Austrian ethologist Karl von Frisch, may represent the most sophisticated example of communication in non-primates. The importance of the discovery was marked by the 1973 Nobel Prize he shared with Konrad Lorenz and Niko Tinbergen.

In 1919, von Frisch noticed that once a dish with sugar-water was discovered by a foraging honeybee, it would be visited by many honeybees from the same hive. He also noticed bees doing a particular waggle dance on the honeycomb in glass-walled observation hives. Von Frisch spent much of the next fifty years on experiments to investigate the phenomenon.

By painting dots of different colours on the bees that visited the sugar-water feeding station, and by watching the marked foragers back at the observation hive, von Frisch was able to study the dances made by foragers coming from a rich food source. Dancing on the vertical honeycomb in the dark, the forager bee makes repeated waggle runs in a particular direction while waggling her body from side to side (Figure 17). The dancer is encircled by bees following her dance movements which give the vector, direction, and distance to the resource (usually a flower patch). The angle to the vertical of the waggle run indicates the flight direction to the food relative to the Sun (Figure 17). The distance to the food source is given by the duration of waggle run. A human observer can now decode the dance. Each second of waggle run duration corresponds to approximately 750 metres (and the arbitrary nature of the code is shown by the way that different races of the honeybee use different run durations to indicate the same distance). Richer resources are signalled by more circuits of the dance.

As the Sun moves clockwise across the sky during the course of the day, the angle of the waggle run indicating a particular food location changes to correspond to the changed location of the Sun. It was this consistent anti-clockwise angle change by 15° per hour that alerted von Frisch to the importance of the Sun compass and indeed how the dance gives the vector direction. If the Sun is obscured by clouds, patterns of polarized light in patches of blue sky give the bees an indication of the position of the Sun. If there is full cloud cover, the bees use landmarks and a sophisticated memory of where the Sun would be at that time of day.

How do the bees estimate the distance to a feeding location? Early studies concluded that the bees gauged distance by the amount of energy consumed to fly to the feeder. However, later studies have shown instead that the bees gauge the distance by measuring how much the image of the world appears to move in their eye during the flight to the food source. To test this, honeybees were trained to enter a feeder by flying through a tunnel just 12 cm wide, where the tunnel walls could be manipulated with different visual stimuli. Bees flying through tunnels with complex visual wall patterns with lots of *optic flow* caused the bees to grossly over estimate the distance they had flown. A 6-metre flight through this tunnel was signalled by waggle dance durations equivalent to flying 180 metres outdoors. Nestmate bees following this dance were misled into flying to a non-existent feeder at the greater distance. Conversely, if the tunnel had black stripes running parallel to the direction of flight, offering an unchanging optical stimulus, the bees' dances indicated a shorter distance.

In the 1960s, some critics argued that the dance was irrelevant and that the information came instead from the odours carried from the location by the dancing bee. The 'bee dance controversy' stimulated ingenious experiments by many researchers which confirmed the key role of the bee waggle dance. These experiments included the *optic flow* tunnel manipulations and other experiments using a computer-controlled mechanical 'bee' programmed to waggle dance and buzz. The mechanical bee's dances could be programmed to send foraging bees wherever the experimenter wished, confirming that humans have understood the code. (Incidentally, synthetic signals from robot 'sage grouse' models, computer-generated frog-calls, and videos of fish have all been used to great effect in studies of communication in those species.)

The responses of individual bees to the dancing foragers allow the colony as a whole to allocate resources to the best food patches over an area of up to 500 km^2, and to prioritize the activity of foragers to the greatest needs of the colony, be they nectar, pollen, or water for cooling the nest. While a dancing bee knows the quality of the patch of flowers she has visited, she does not know the value of her resource to the colony at that moment. The answer comes from how quickly 'unloader' bees take her cropload of nectar. If her load is taken quickly, nectar is in demand and she will do a large number of waggle dances to the patch she has found. If she has to queue to be unloaded, she may not dance at all.

The waggle dance is also used by scout bees to communicate the location of potential nests that the swarm could choose as their next home (Chapter 7).

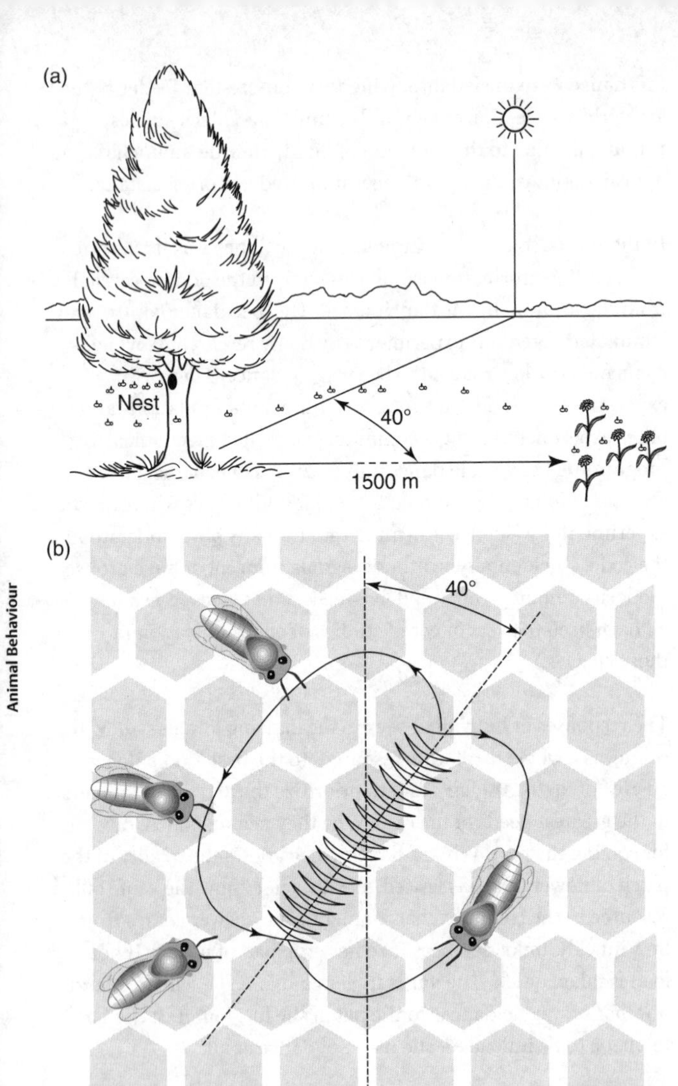

17. A honeybee forager gives fellow workers the direction to rewarding food sources (a) by the angle of the waggle section away from the vertical, which is the angle between the path to the food and the Sun's azimuth (the point on the horizon directly below the Sun (a, b)). The duration of waggle (central part of the dance) codes the distance to the food (c).

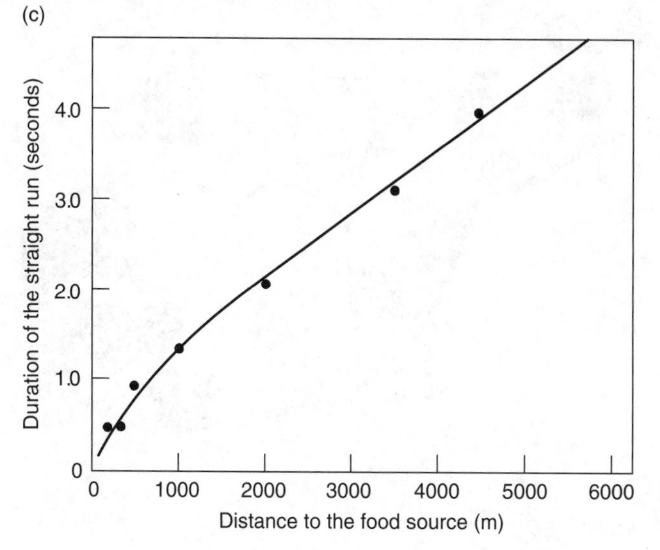

17. Continued

Vervet monkey alarm calls

Vervet monkeys (*Chlorocebus pygerythrus*) are a group-living species found in savannah regions of East Africa. If a vervet monkey spots a dangerous predator, it gives a different-sounding alarm call depending on whether it is a terrestrial predator such as a leopard, an aerial predator such as an eagle, or a snake such as a cobra (Figure 18). Other nearby vervets respond to these calls in different, apparently highly adaptive ways appropriate to the danger posed by that predator. Vervets on the ground respond to the loud barking of the leopard alarm call by running into the tree tops, where leopards cannot follow. Vervets respond to the double-syllable 'cough' eagle alarm by looking up, then running for cover in a bush. On hearing the 'chutter' snake alarm, vervets respond by standing up on their hind legs and scanning the grass.

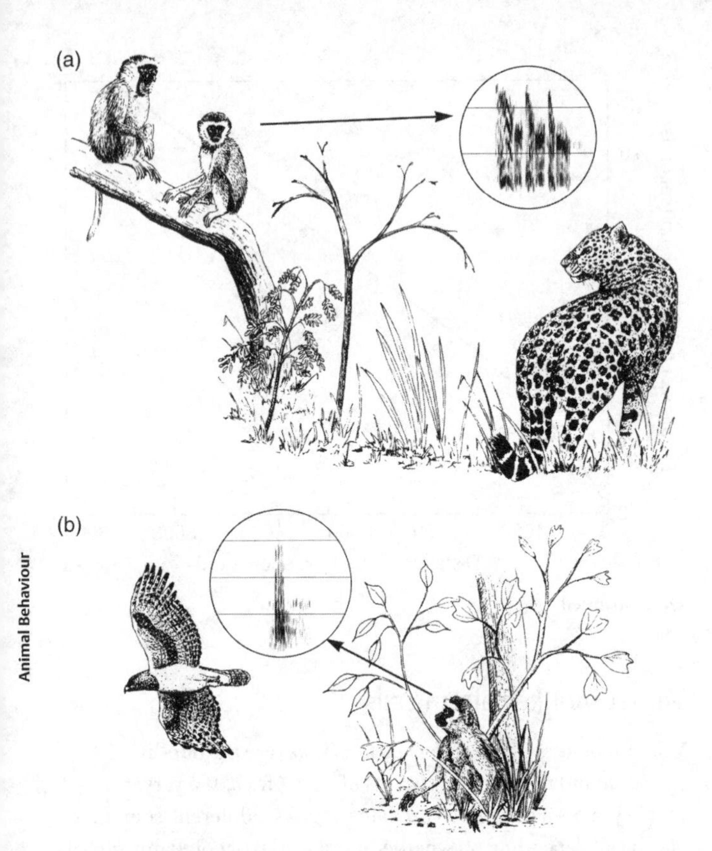

18. Vervet monkeys make different alarm calls depending on the danger they detect. The behaviour of the receivers matches the calls. Leopard (a); eagle (b).

Dorothy Cheney and Robert Seyfarth recorded alarm calls made by members of the group on seeing different predators. Then on another occasion, when no predator was present, they played back the calls from hidden loudspeakers. The vervets responded appropriately to each of the playback alarm calls even in the absence of the predator. For example, they scanned the sky and ran to a bush after an eagle alarm, showing that the adaptive behaviours were in response to the call alone.

Young vervet monkeys often make 'mistakes' when they first start to make alarm calls: they may give alarms when they detect species like warthogs and pigeons that pose no danger. However, there is a pattern to their mistakes. Young vervets make the eagle alarm only when they detect birds and other things in the air, such as a falling leaf. Leopard calls were made only in response to terrestrial mammals, and snake calls only to long snake-like objects. How do the infants eventually learn the correct association between a particular alarm call and, for example, a dangerous eagle species? If an infant is the first to make an eagle alarm, nearby adults still look up. If it is a false alarm, the adults return to what they were doing. However, if the infant has spotted a dangerous eagle species, the adults are likely to give an alarm call themselves. Cheney and Seyfarth speculate that this may reinforce correct calls by the infant.

The vervet alarm calls have become a textbook example of non-human communication of information and the function of signals. However, unlike the honeybee dance which has been extensively researched by many scientists worldwide, the vervet alarm call studies had not been repeated until recently. The repeat study, on a different population of vervet monkeys in South Africa, did not show the clear-cut adaptive responses found in the original studies from the 1980s. A number of reasons could explain these results, including larger vervet group sizes in the replication in 2014. This highlights a problem in animal behaviour: there is rarely the opportunity for replication by independent researchers. What usually happens is that similar behaviour is investigated in new species. For example, in the case of alarm calls, many animal species, including ground squirrels, chickens, and forest-dwelling primates, have been shown to have different alarm calls for different kinds of predators.

This question of replicability or reproducibility is a problem for the whole of science, not just animal behaviour. A 2015 project which repeated a hundred influential studies in psychology found that a worryingly small percentage (less than half the studies) gave the same findings. The reasons are largely to do with small sample sizes combined with a desire of scientists to find and publish novel positive results. Psychologists are actively addressing the problem. Animal behaviour researchers have some catching up to do.

Interspecies communication

Vervet monkeys also learn to respond appropriately to the predator-specific alarm calls of other species, including birds called superb starlings, which share their habitat. If animals have predators in common, responding to the warnings made by other species is good for survival.

Interspecies communication enables cooperative hunting by different fish species with complementary hunting skills. In the Red Sea, coral reef groupers (*Plectropomus pessuliferus*) use a repeated body-shaking 'shimmy' signal to invite giant moray eels (*Gymnothorax javanicus*) to join them for a mutually beneficial hunting trip. The moray eels can hunt prey in coral crevices that are out of reach of the groupers, and the groupers, which are open water predators, can catch fish flushed out by the moray. The groupers sometimes use a 'head-stand' display, a *referential signal*, to point where prey are hiding. Both partners benefit from higher prey capture rates.

In Mozambique, human honey-hunters follow the greater honeyguide (*Indicator indicator*), a bird which flies ahead of them, leading them to nests of the wild honeybee in large trees in a rare example of a mutualistic foraging partnership between humans and free-living wild animals (Figure 19). Both sides of the partnership make signals. Greater honeyguides seeking a

19. Yao honey-hunter Orlando Yassene holds a wild greater honeyguide (temporarily captured for research) in the Niassa National Reserve, Mozambique.

human collaborator, approach people and give a loud chattering 'tirr-tirr-tirr-tirr' call. This call is distinct from their territorial song and is accompanied by referential gestures: the bird flies from tree to tree, flashing its white outer tail feathers, in the direction of the bees' nest until its human follower finds the nest. The honey-hunters also make calls. Playback experiments showed that a specialized vocal sound made by Mozambican honey-hunters seeking bees' nests doubled the probability of being guided by a honeyguide from about 33 to 66 per cent. This increased the overall probability of finding a bees' nest from 17 to 54 per cent, as compared with other animal or human sounds of similar amplitude. The bird benefits from the mutualism by gaining access to feed on the honeycomb of the bee nest that it could not otherwise access. People in other parts of Africa including Tanzania and Kenya also follow the greater honeyguide but the honey-hunters' traditional calls are different in each place.

Exploitation of signals: eavesdropping and deception

Broadcast signals can be eavesdropped by other species—to devastating effect. For example, male crickets singing to attract a female also attract a parasitoid fly which will kill them (Chapter 1). Another well-studied example is the tropical Túngara frog (*Physalaemus pustulosus*) in Panama. If a male is alone, he sings to attract females with a simple call (whines only). However, if he is joined by other males in a chorus, the male adds up to seven 'chuck' sounds to make a complex call of whines and chucks, which the females prefer and find easier to locate. One evening, a frog researcher, Michael Ryan, was helping a colleague to mist-net bats nearby only to discover that many of the bats had Túngara frogs in their mouths. Playback experiments confirmed that the specialist frog-eating bat (*Trachops cirrhosus*) homes in on the frog call—a loudspeaker playing the frog call has them swooping in. Like the female frog, the bat also finds the chuck call easier to locate. The bat has evolved specializations of its hearing which allow it to be sensitive to both its own ultrasonic echo-location calls (>50,000 Hz) and to the much lower frequency calls of the frogs (<5,000 Hz). The male frogs' only defence is to stop calling when they detect a bat swooping in towards the pond. If being eaten by a bat was not bad enough, the frog males' songs also attract specialist blood-sucking flies, which also prefer the complex calls.

Communication is also risky when a signal appears to be from a potential mate, but is instead being made by a deceptive predator. The night-time light-flash signalling system of bioluminescent fireflies (beetles in the family Lampyridae) is one that has been exploited in this way. Hovering *Photinus* firefly males flash a species-specific 'flash-code' of light-flashes from the lantern of light-producing tissue in their abdomens. When a female of the species responds with her flash reply, leading to a light-signal duet, the duet culminates in the male locating the female and courting her. The danger to *Photinus* males comes from females of another firefly genus, *Photuris*, which have evolved to be specialist predators. Until she has mated herself, the *Photuris* female responds only to the light-flashes given by males of her own species, but once mated her behaviour changes. She becomes a 'femme fatale', giving deceptive light-flashes in response to the flashes from *Photinus* males as if she was a *Photinus* female. The *Photinus* males are lured in by the duet and, once the male *Photinus* is near, the *Photuris* female pounces and eats it. Field experiments have shown that individual *Photuris* females can respond with flash-codes to match the codes of different *Photinus* species. Over evolutionary time, the risk of predation by *Photuris* may have shifted the flash-codes and the evening hours that *Photinus* species use.

Another form of deception is used by female bolas spiders, which do not spin orb webs to catch insects. Instead, the female puts the silk into a sticky *bolas* (ball) and loads it with counterfeit female moth pheromone. The pheromone lures the male moth within reach and the spider swings its sticky bolas, usually catching the moth. Some species of bolas spider can produce one combination of molecules to attract males of one moth species early in the evening and switch production to that of another moth species when that moth becomes active later in the evening.

Chapter 6
Winning strategies

Behavioural ecology

The field of behavioural ecology, developed in the 1960s and 1970s, offered new ideas and provided powerful ways of exploring how behaviour evolves. Behavioural ecology examines how the evolution of behaviour is related to an individual's chance of survival or reproductive success through specific questions, such as how long a bird should look for food in a particular patch. Behavioural ecology also uses comparative approaches, asking questions such as: Why do both sexes look after the young in some species whereas, in other species, only one sex is involved? Why do chimpanzee males have such large testes for their body size compared with gorilla males?

Adaptation has been a key unifying concept for behavioural ecology, leading to the testable expectation that behaviours should evolve to maximize the 'fitness' of individuals showing those behaviours. Fitness is a measure of an individual's success in passing on copies of its genes to future generations. Because it is often hard to measure lifetime fitness, researchers have instead used short-term measures such as energy intake or success in escaping predators.

Earlier generations of researchers tended to ignore individual variation in their search for characteristic species-wide behaviours. Behavioural ecologists, however, realized that by carefully measuring the variation in behaviour between individuals, adaptive hypotheses could be tested by exploring the effects of these differences in behaviour on survival and reproduction. For example, behavioural ecologists might ask: Which parents would leave more offspring: those laying four eggs or those laying five?

The idea that adaptation should lead to selection for maximization of fitness has led to optimality models, which offer quantitative predictions of how animals should behave. For example, food choices may be explained by seeing which prey are most energetically profitable.

Economic 'decisions'

A study of crows (*Corvus caurinus*) dropping whelks (marine snails with thick shells) onto seashore rocks to break them open has become a classic example of the economics of foraging. Reto Zach observed the crows on the west coast of Canada and noted that the birds selected only large whelks, dropping them from an average height of about 5 metres. What could explain these behaviours? To find out, Zach set up a 15-metre pole with a movable platform that would allow him to drop shells from different heights, like the crows. He found that small whelks took many more drops to break than large whelks (and small whelks had much less snail flesh inside). When Zach calculated the energetic cost of a crow's flights and compared it to the energy gained from eating a whelk, only large whelks would be profitable, making them a sensible choice. Why drop from 5 metres? Zach reasoned that upward flight was energetically expensive. He found that whelks dropped from 2 metres took fifty drops on average to break, but when dropped from 5 metres, the shells were likely to break in just four drops. Thus, breaking a whelk by dropping from a height of about 5 metres required the least total upward flight, and this was very close to what Zach observed the crows doing. This finding suggests that crows try to minimize the energetically expensive upward flight. In other words, they behaved in ways that maximized their energy intake.

Decision-making also plays a role in reproduction, as mating animals must make decisions about how long to mate. For male yellow dung flies, the currency is time rather than energy. The longer a male mates with a female, the more of his sperm will replace those of her previous partners. Geoff Parker showed that while the rate of sperm replacement is high at first, there are diminishing returns the longer the male stays mating with the female, and at a certain point the male might do better by moving on. When Parker calculated the costs and benefits, he predicted an optimum mating time of forty-one minutes to gain the maximum number of eggs fertilized per minute. This prediction was close to the observed average copulation time of thirty-six minutes. However, a major problem with optimality models is knowing when 'close' is close enough to make the proposed model a reasonable explanation of the observed behaviour.

The use of quantitative analysis and models to predict and explain behaviour can highlight discrepancies between our predictions and the animal's observed behaviour. These discrepancies may suggest that we are not measuring the currency that the animals are using. For example, birds feeding their nestlings seem to maximize the rate of energy delivery to the nest (the more grubs per minute, into the mouths of their nestlings, the better), rather than maximizing energetic efficiency when they are foraging for grubs. Honeybees, by contrast, seem to maximize energetic efficiency when they are foraging for nectar.

Models assume that animals are well-designed problem-solvers, but of course animals do not make calculations in the same way as the human behavioural ecologist observing them. Instead, it seems that animals have evolved to follow rules of thumb that give more or less the right answer. For example, it seems that honeybees may use some measure of the total weight they have carried during their flight as their rule of thumb. When researchers added tiny weights to the backs of honeybees as they foraged, the bees returned to the nest with a smaller load than unmanipulated control-group bees.

Why do animals sometimes appear to behave less optimally than researchers might expect? One reason is that their behaviour may be a trade-off between competing pressures, such as the need for food set against the risk of being eaten. Hungry stickleback fish (*Gasterosteus aculeatus*), given a choice of a low and a high density of prey (*Daphnia* water fleas), prefer the higher density. However, if a model of a predator (a kingfisher bird, for example) is flown over their fish-tank (Figure 20), the sticklebacks change

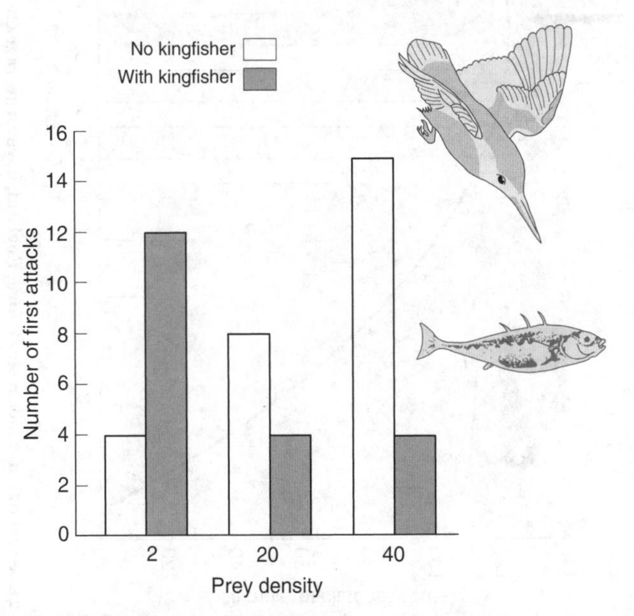

20. Animals balance risk and reward. Sticklebacks usually choose to attack high-density areas of prey, but after a model kingfisher was flown over the tank they chose low-density areas.

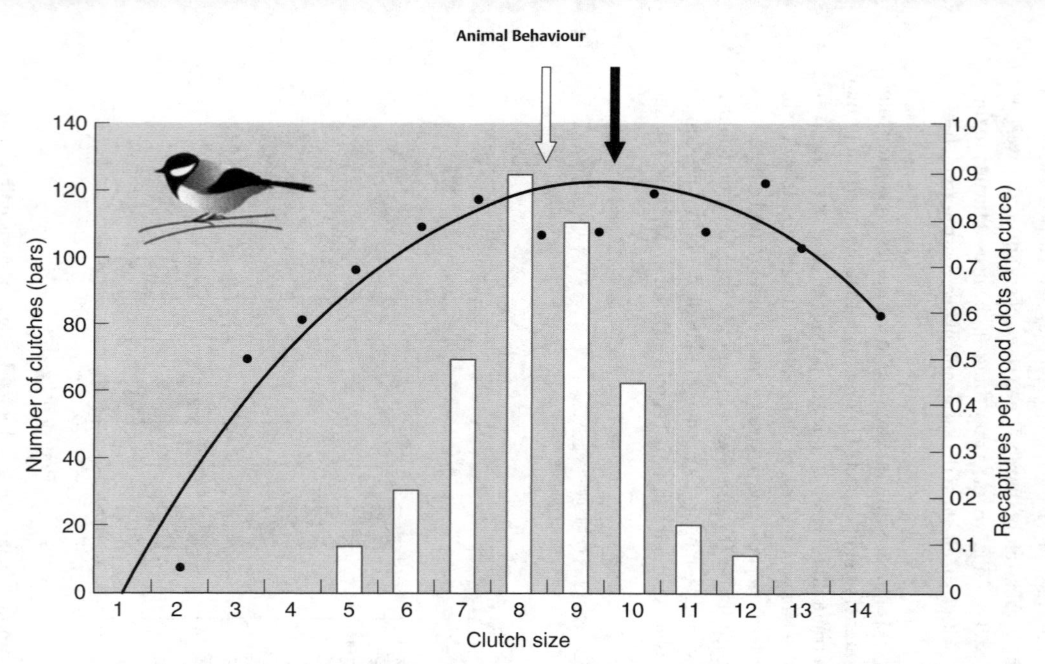

21. Great tits in woods near Oxford, England, have an average clutch of about 8–9 eggs (bars) (white arrow). When eggs were added or removed in a manipulation experiment, it was discovered that a pair of great tits could rear a larger number successfully (shown by recaptures, dots and curve, black arrow).

their feeding behaviour and instead choose the low-density patches of *Daphnia*. The trade-off is between feeding rate and the greater vigilance possible when feeding on low-density prey.

Another explanation for behaviour which seems sub-optimal at first sight is that there might be trade-offs between short-term benefits and longer term outcomes, such as the chance to breed in future years. This was explored in a classic study by Chris Perrins on the great tit (*Parus major*), a small bird which lives up to two years. The optimum number of eggs to lay in a clutch is the number that will give the most offspring surviving to breed; too few eggs in the clutch means that all will survive but they will be few in number, while laying too many eggs in the clutch may result in young birds that are too small to survive the first winter because the parents' ability to bring enough food is limited (Figure 21). This is a quality–quantity trade-off. When Perrins added eggs to nests in an Oxford wood, it turned out that the greatest number of surviving offspring came from a clutch size that had almost two more eggs than the average clutch size observed in the wild. Why did the birds lay fewer than this higher number? The likely answer is that this short term 'optimum' was based on only one season's breeding, whereas the extra feeding effort for extra offspring reduces the chance the female will survive the winter to breed again next year. Natural selection has maximized *lifetime* reproductive output rather than maximizing the size of each clutch at the expense of future chances to breed.

Competing for resources

In the same way that we judge which queue to join in a supermarket, the best foraging and mate-searching strategies may depend on what other individuals are doing. In one example, researchers tested this idea by counting ducks on a pond when two people, at different ends of the pond, threw in food at different rates. In eighty seconds, the ducks sorted themselves out to match the rewards: if twice as much food was being thrown in at one end, twice as many ducks moved there. If the difference in reward rates was switched, the ducks quickly moved to the other end. This is something you can try yourself with some ducks and a friend—it works.

If they cannot successfully compete directly, some animals follow alternative strategies to find mates. Female natterjack toads (*Bufo calamita*) are attracted to the males making the loudest calls in a chorus. Smaller males cannot compete with the loudness of the calls made by larger males, so small males adopt a 'satellite' behaviour, sitting silently near the calling males and attempting to intercept the attracted females. The small males are making the 'best of a bad job' until they grow larger themselves.

Parental care and mating systems

One of the successes of behavioural ecology has been to explain the pattern of parental care across species: Why in some animal species do both parents look after their young while in others just one sex (usually the female) provides parental care? And why in still other species, is there no parental care? These different parenting patterns, combined with an understanding of a species' ecology, also help to explain the differences in mating systems across the animal kingdom—from species with monogamous pair-bonding to other species with harems, to species with no pair bonding (so-called promiscuous mating). There are patterns across different animal taxa. For example, care by both parents is common in birds. Most mammals look after their young, but in most species only the female is involved. For most invertebrates there is no parental care of offspring.

However, there are exceptions to the generalizations. For example, some species of invertebrate do show parental care. One is the burying beetle *Nicrophorus*, where parents can improve survival of the young by feeding or protecting them. After discovering a dead mouse, the male and female bury it in an underground nest chamber. They fashion the mouse flesh into a nest shape and the eggs are laid on this. The parents feed their beetle grubs with regurgitated flesh when they beg, just as bird parents do.

In many bird species, successful rearing of chicks is limited by the rate that food can be brought to the nest. If two parents can feed twice as many young as a single parent (as is the case with a forest species of weaver bird that feeds on insects), then both male and female will increase their reproductive success if they stay together. For either parent, deserting the nest would halve the surviving brood and they would still have to find another partner. In such species, there is parental care by both parents.

If food is so abundant that a single parent can supply the nest, then one parent is likely to desert. Usually this is the male, as he can increase his lifetime reproductive success by mating with other females, whereas a female's reproductive success is limited by the number of eggs she can produce. Weaver bird species living in the savannah and feeding on plentiful seeds are polygynous (each male mates with several females) and only the females look after their broods.

In mammals, the male cannot contribute during gestation (apart from feeding and protecting the female), and only females produce milk for the newborn young. Thus, it is perhaps not surprising that in most mammal species, parental care is provided by the female alone, and males desert to mate with other females. Carnivores such as foxes, however, are among the few exceptions where the male can contribute by bringing meat to the offspring once they are weaned.

By contrast, in the small proportion of fish species with parental care, it is the male that usually provides the care. The stickleback, studied by Niko Tinbergen, is a typical example. The male courts females to his territory on which he has built a nest. After each female has laid her eggs, he follows her into the nest, releasing his sperm to fertilize the eggs. He will spend the next two weeks fanning the eggs, to keep them well oxygenated, and chasing away predators. Once the eggs hatch, he will shepherd the young. Why male parental care in fish? The phenomenon seems to be associated with external fertilization and the importance of holding a territory. While the original function of defending a territory may have been to attract more females, there are few additional costs to looking after the eggs, and eggs from many females add more reproductive success for the male without adding costs. Some fish species take it further, with the male brooding the eggs in his mouth or in a special pouch as in sea horses or pipefish.

Who looks after the young, if at all, helps explain mating systems. In species needing both parents to feed the young, monogamy is common, such as in many bird species. Where the male does not contribute to raising offspring, and where food is plentiful enough for females to live within his territory, mating systems may evolve in which a male defends a harem of females. This is the mating system of red deer and gorillas. In both of these species, there is sexual selection for large size and/or weapons such as antlers in the males, as they fight each other to be 'king'.

In a very few bird species, males provide all the parental care. One example is the South American water bird, the wattled jacana (*Jacana jacana*), which lives in food-rich swamps. Because females can produce more eggs than they can brood, they fight with each other to control a territory containing up to four males. The males brood the female's eggs and look after the young. Sexual selection has selected for increased size in the competing sex—in this case, the females—which are more than 50 per cent heavier than males.

The sex role reversal in the jacana, with females that compete for access to males which will incubate the eggs, is a powerful support for the explanations provided by behavioural ecology. This, and other exceptions to the 'rule' of female parental care in birds, is explained by the ecology of the species.

Sex ratios

We seem to take for granted that there are equal numbers of males and females in every species. However, in recent years it has become clear that in many species, from parasitic wasps to red deer, females can influence the proportions of males and females in their offspring. Why then don't mothers maximize their number of potential grand-offspring by producing many daughters but just one son (which could mate with many females)? The British statistician R.A. Fisher showed logically that the proportions of male and female offspring would always return to equality as at equilibrium the average fitness of a male must equal that of a female. This is because if offspring of one sex had an advantage by being rarer, other females would be selected to produce offspring of that sex until balance was reached again. The equilibrium can be in the numbers or investment in each sex. If, for example, males take twice as much energy to raise, then the equilibrium will be to produce twice as many female offspring as males. An exception to the equilibrium, which shows the logic in action, comes from species in which brothers tend to mate with their sisters, such as some fig wasps. In these species, the mother produces an almost all-female brood, with just enough sons to fertilize her daughters inside the fig fruit—the daughters will not leave the fruit before being fertilized. This skewed sex ratio can evolve because the sons of other mothers have little access to the fig fruit.

Sexual selection

The male peacock's extravagant tail might seem to be a challenge to natural selection as it surely makes him more vulnerable to predators. Charles Darwin proposed that a kind of sexual selection—that of female choice—might explain how brightly coloured males with fabulous feathers or elaborate songs (or powerful smells, as in male goats) could evolve despite their likely cost to the survival of the males.

He also noted another kind of sexual selection in species such as red deer in which males fight each other to gain access to females. In these species, selection on males for superior fighting ability leads to the evolution of larger male body size and weapons such as antlers and tusks. (Where females are the competing sex, as in the wattled jacana, there is selection for greater size and fighting ability in that sex). Sexual selection is a kind of natural selection that results from differences in reproductive success rather than from differences in survival.

Examples of female choice have been harder to study than male–male combat. However, a neat experiment showed that female long-tailed widowbirds (*Euplectes progne*) do choose males with longer tails. Malte Andersson cut short the tail feathers of some males and superglued those cut feathers on other males to extend their tails. Control males had their tails cut and glued back at the same length. More females chose to nest in the territories of the males with the elongated tails (Figure 22). Similarly, male European sedge warblers (*Acrocephalus schoenobaenus*) with more complex songs were the first to acquire mates in the spring. Clive Catchpole and colleagues also showed that females were more responsive to more complex male songs in the laboratory too.

How many sexual partners do animals have? Almost 90 per cent of bird species appear to be monogamous, with a male and female pairing on a territory to rear their young. However, a real surprise has come from DNA fingerprinting, which has revealed that in many socially 'monogamous' bird species the young birds in the nest have been fathered by multiple males. In black-capped chickadees (*Poecile atricapillus*), females actively search out these extra-pair males. Infidelity by both sexes seems to be common in animals of all kinds. Mating with more than one partner leads to the possibility of sperm competition.

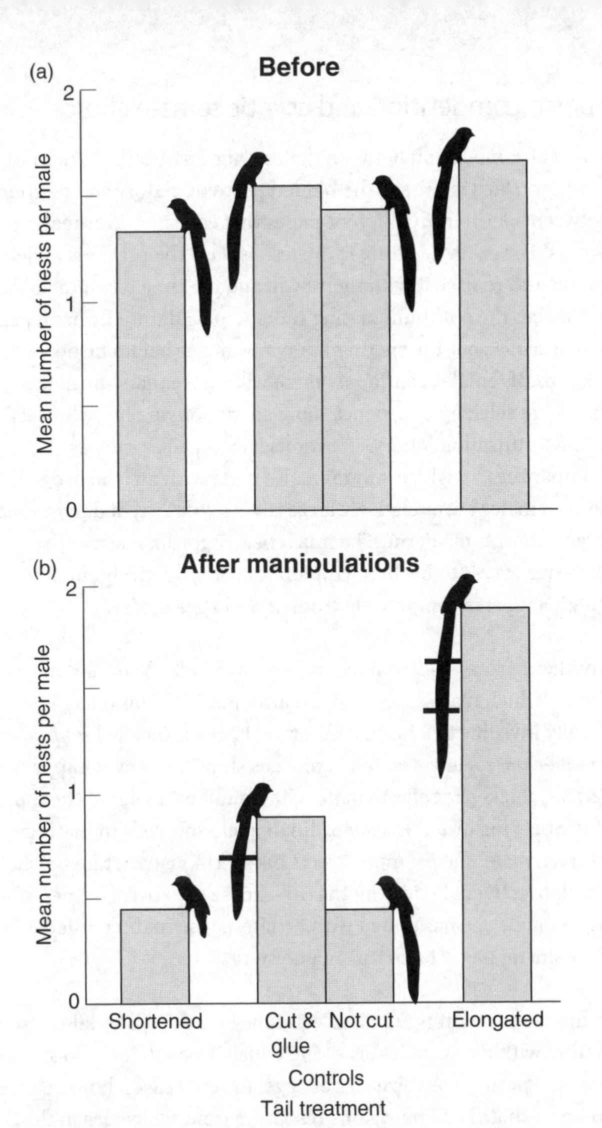

22. Male widowbirds with longer tails are more attractive to females. Before the experimental manipulations, all the males were equally attractive (a). After some males had their tails shortened or lengthened (b), females nested most in the territories of males with the super-elongated tails.

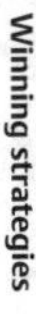

Sperm competition and cryptic female choice

The real competition between males is for fertilizations, not just matings. Darwin missed the hidden post-copulatory competition between sperm from different males that can start after mating occurs. For example, dung fly males displace the previous male's sperm and replace it with their own, and the time taken to do this influences the optimum mating time. A male damselfly has a penis which scoops out the sperm of previous males before he deposits his own. If females commonly mate with more than one male, males are selected to produce more sperm, buying more 'tickets' in the fertilization lottery. In promiscuous species such as chimpanzees, in which several males mate with an ovulating female, males have relatively large testis (120g) needed to produce large amounts of sperm. The much bigger gorilla male, with exclusive access to his harem of females, has relatively small testis (30g), as sperm competition from other males is rare.

Another surprise is that females in many kinds of animal can choose which males' sperm to use after mating. This effect is usually invisible to a human observer, hence it is called *cryptic female-choice*. However, feral chickens show one way it happens. Female chickens prefer to mate with dominant males at the top of the pecking order. If a subordinate male succeeds in mating with a female, she promptly ejects his sperm just after he finishes copulating. Female tiger moths can choose the sperm of the largest male she has mated with, whatever the mating order. How she manages to do this is a mystery.

In many species, it is not clear what the benefit to a female is from mating with many males, as a single male's ejaculate could supply enough sperm to fertilize all her eggs. In some cases, however, we do know that the female gains resources from molecules in the ejaculate. For example, in the case of tiger moths, along with the sperm in the ejaculate is a gift of poisonous plant molecules (which will protect her and the eggs from predatory insects and spiders) and nutrients (which increase her egg output by about 15 per cent with each mating).

Sexual conflict

A gene's-eye-view has helped explain why, contrary to what one might expect, the individual interests of females and males may be in conflict. For example, the male fruitfly *Drosophila* includes protein molecules in his ejaculate that hijack the female's nervous system, making her unreceptive to the courtship from other males. The male seminal proteins that have these effects have been selected for because they reduce the probability of sperm competition. However, the resultant change in the female's behaviour may not be in her interest, as she might benefit from mating again (for example, to ensure that she has fertile sperm in case the first male's sperm were faulty). As a result there is an ongoing 'arms-race' over evolutionary time: females are selected to evolve to be unaffected by the seminal proteins (by selection for receptors no longer sensitive to them), while males' seminal proteins undergo new rounds of selection to regain the manipulative effect.

Explaining altruism

It is not surprising that parents look after their own offspring. It is surprising, however, when animals that do not produce offspring of their own help others to reproduce. This is called altruistic behaviour, illustrated by sterile honeybee workers helping their mother, the queen, produce more sisters. Charles Darwin recognized this was a problem. How could helping behaviour evolve if the workers leave no offspring of their own? Darwin realized the answer was family. Just over a hundred years later, Bill Hamilton showed how copies of the workers' genes (including the 'genes for helping') get passed on indirectly through relatives. For honeybee workers, these relatives are their sisters which become queens.

Hamilton's inclusive fitness theory explains how genes in a helping individual can be carried to the next generation as long as the help allows parents or other blood relatives to produce sufficient extra offspring (*indirect fitness*) to make up for the cost to the helper's own direct reproduction (*direct fitness*). For animals helping their parents or relatives but not reproducing themselves, their inclusive fitness will be via indirect fitness in the form of the reproductive success of the animals they helped to rear. The process of helping relatives is often called *kin selection*. Hamilton's ideas were popularized by Richard Dawkins in his book *The Selfish Gene*.

Social behaviour will evolve under specific combinations of relatedness, benefit, and cost. Helping relatives is favoured by natural selection if the benefit (B) to blood relatives (carrying the gene by shared descent), discounted by how closely related they are (r), is greater than the cost (C) in direct fitness lost by the helper. This gives Hamilton's rule: altruistic behaviour is favoured if $rB > C$. Hamilton's rule predicts that positive relatedness is a necessary condition for the evolution of altruism, and altruism evolves more readily when B is high and C is low. The greater the relatedness, the greater the likelihood of altruistic behaviour evolving.

Your parents each contributed one of your two copies of each gene, so your r to each parent is half or ($r = 0.5$). Similar logic can be used to work out the probability of sharing a gene copy in near and far relatives: between brothers and sisters, $r = 0.5$; and r between an individual and a cousin is one-eighth = 0.125. Hamilton's rule means that helping even distant relatives such as cousins could be selected for so long as enough of them were helped to reproduce. Thus, it would benefit an altruistic individual to give up one offspring of their own ($1 \times r = 1 \times 0.5 = 0.5$ genetic units) to save five cousins ($5 \times r = 5 \times 0.125 = 0.625$ genetic units). There's a story, probably apocryphal, that having quickly done the calculations on an envelope, the 1930s British biologist J.B.S.

Haldane said he would happily lay down his life to save two brothers or eight cousins.

Favouring relatives

Many kinds of animal, perhaps most, are not able to recognize relatives. Instead, animals seem to use rules of thumb, which could be as simple as, 'treat young animals in your nest as your offspring'—as this assumption will usually be true. However, animals without nests must learn to recognize their young. Bat parents learn the calls of their offspring so that in a crowded bat cave they can feed their young and not non-relatives. Mother sheep similarly learn the smell, face, and calls made by their lambs and don't feed others. Another way to recognize relatives is *self-matching*—comparing yourself to them. Richard Dawkins called this the 'armpit effect': sniff the stranger and your armpit. If the stranger smells like you, they are likely to be kin. Belding's ground squirrels use this to recognize siblings when they meet again after hibernation.

Cooperative breeding

In about 10 per cent of bird species, young birds stay with their parents to help for at least one breeding season before attempting to breed themselves. This is called *cooperative breeding*. Why do the helpers help? The reason seems to be the indirect reproductive fitness they gain by helping their parents rear related siblings. The help may take the form of bringing extra food for the nestlings or defending them from predators. Each helper in the white-fronted bee-eater (*Merops bullockoides*) in Kenya had a major effect in increasing nestling survival, on average enabling the parents to raise an extra half chick. Because the helpers were closely related to the nestlings they helped to rear (average $r = 0.33$), they gained a large indirect fitness through increasing the production of these relatives. Against these benefits is the low cost to helpers of forgoing producing their own offspring. This is because any helpers which tried to breed as pairs by themselves would struggle to rear any offspring successfully in the unpredictable and harsh conditions of their habitat. Removing the helpers in another bird species, the Florida scrub jay (*Aphelocoma coerulescens*), more than halved the number of fledglings successfully produced by breeding parents.

The meerkat, a social mongoose, lives in the Kalahari Desert in family groups of up to fifty individuals. Only the dominant female and male reproduce in this cooperatively breeding species. Their offspring help to rear younger siblings (average $r = 0.3$) by feeding them scorpions, lizards, and other prey. The feeding by helpers increases the growth and survival of the pups and more than doubles the chance that these pups will compete successfully to become dominant individuals that are able to reproduce, all adding to the indirect fitness of the helpers. Helpers also gain direct fitness by helping, as the larger the group, the better their own survival. Even though helpers give up direct reproduction, the costs are low, as a pair of helpers leaving the group cannot reproduce successfully.

Evolution of eusociality

In cooperatively breeding birds and mammals, the helpers may themselves get the chance to reproduce later. In *eusocial* species, there is a reproductive division of labour, with a specialized worker caste that is effectively or completely sterile. Examples of eusocial animals include the naked mole-rat, some social coral reef shrimps, and the more familiar eusocial ants, bees and wasps, and termites. In these eusocial species the workers' fitness is indirect: the workers help their mother, the queen, rear other workers, ultimately rearing a few sisters and brothers which will become future queens and reproductive males.

Hamilton's rule predicts that high r favours the evolution of altruism, in particular for the highly developed form shown by eusocial species. Eusociality has evolved independently at least eight times in the Hymenoptera (ants, bees, and wasps). It has also evolved independently multiple times in the termites (which are a specialized kind of cockroach). Monogamy of the founding male and female pair seems to be an essential ancestral feature for the evolution of eusociality across the animal kingdom, as it ensures the highest relatedness for the worker offspring who are thus on average as related (0.5) to their brothers and sisters as they would be to their own offspring (0.5).

Naked mole-rat (*Heterocephalus glaber*) colonies live underground in tunnels that offer protection from the harsh climate of arid parts of East Africa. Only the queen gives birth, supported by up to 300 of her offspring which rear the young and dig tunnels to find roots and tubers to feed the colony (Figure 23). The cost (*C*) for workers of forgoing their own reproduction is small because a single naked mole-rat could not survive on its own.

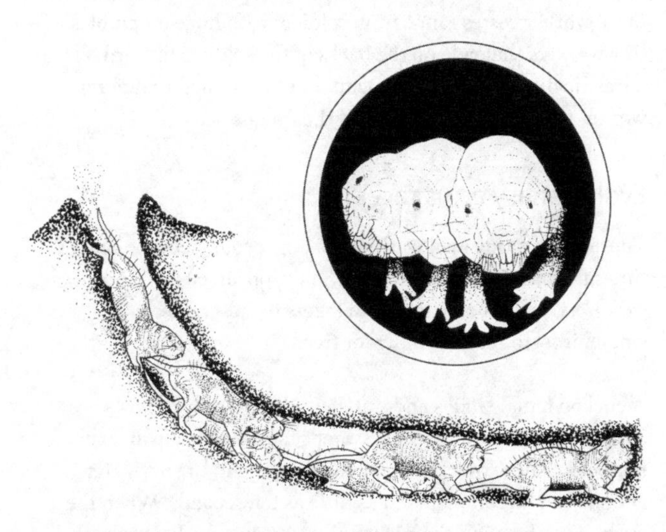

23. Naked mole rats are eusocial mammals with a queen and workers, analogous to social insects. Here workers are shown digging a tunnel together.

In naked mole-rats, and in social insect species with small colonies of perhaps tens of workers, the queen signals her dominance to the workers by physical 'bullying'. In social insects with large colonies of up to a million workers or more, the queen instead signals her presence and fertility with a 'queen pheromone' which spreads throughout the colony, passed from worker to worker. If the queen pheromone is present, workers' ovaries remain undeveloped and they do not attempt to reproduce, instead rearing more sisters. The queen pheromone is an honest signal (Chapter 5) that the queen produces only if she is fertile and laying eggs. Evidence is growing that queen pheromones in the eusocial ants, bees, and wasps have evolved independently, from pheromones that signalled fertility in their respective solitary ancestors.

In many ant and termite species there is great specialization of physically different worker castes. In leaf-cutter ants, for example, these worker castes range from soldiers with huge mandibles to tiny workers that ride on the backs of the soldiers to ward off parasitic flies. Inside the leaf-cutter ant nest, further different worker types tend the brood and the fungus beds.

Evolutionary arms races

The sight of a tiny reed warbler feeding a young cuckoo ten times its mass is unforgettable. The European cuckoo (*Cuculus canorous*) is a brood parasite that gets its hosts, species of small birds, to rear its young for free.

A cuckoo female hides and watches until the parent hosts are away from the nest, then she swoops in and picks up an egg in her beak, before quickly laying her own egg in the nest. Her actions can be completed in as little as ten seconds. When the cuckoo egg hatches, the cuckoo chick pushes the host eggs and any host nestlings out of the nest. The single cuckoo chick then manipulates the hosts to feed it generously by making a rapid begging call ('si, si, si, si...'), which sounds just like a whole brood of host chicks.

The hosts lose all their own offspring so, in this evolutionary arms race, there is strong selection for adaptations against the cuckoo's parasitism and counter-adaptations from the cuckoo to evade these defences. In response to cuckoo parasitism, host species have evolved a new behaviour of rejecting eggs that do not match their own. In response to egg rejection by hosts, cuckoos have evolved eggs that beautifully mimic the colouration and spots of their host species' eggs (this can evolve because the cuckoo has separate races that each specialize on parasitizing a particular host species). In response to this egg mimicry by the cuckoo, host females have been selected to have an individual colour pattern 'signature' on their eggs, making the cuckoo egg stand out as a 'forgery'.

Host species have been selected to respond to the presence of the female cuckoo by gathering together to mob it. However, the cuckoo, in response, has evolved hawk-like plumage (which deters mobbing) as well as many different colour morphs that make it harder for host birds to learn to recognize it.

Ingenious field experiments by Nick Davies and others have explained much about the ongoing arms races between cuckoos and hosts. Hosts must avoid the costs of misidentification and rejecting their own eggs or young by mistake. One mystery is why the hosts of the European cuckoo do not reject the cuckoo chick once hatched, whereas the Australian host bird species of the bronze cuckoo (*Chalcites* species) do reject cuckoo chicks.

The future of behavioural ecology

The successes of behavioural ecology have been many. A real strength has been the way that kin selection theory, for example, has produced testable hypotheses. Especially notable to me are the explanations for the evolution of eusociality, mating systems, and the roles of the sexes in parental care. Optimality and modelling approaches have been helpful in explaining foraging behaviour and sex ratios.

However, behind much of behavioural ecology there is an untested assumption that there is a heritable genetic basis underlying the behavioural traits being studied. The lack of genetic evidence underpinning these traits is not surprising as traditional genetic analysis of complex multigene systems in wild animals has simply not been possible. However, the new genomics technologies are bringing the cost of DNA sequencing down low enough to sequence wild species whose behaviour is being studied, not just model animals such as the lab mouse and *Drosophila*. For example, genomics allowed the mutations in the silenced crickets (Chapter 1) on the two islands to be compared—and be shown to have occurred independently. The behaviour of the ruff (*Philomachus pugnax*), a Palearctic wading bird, has been much studied for its spectacular mating system, with three different male morphs that differ in sexual behaviour, body size, and plumage colour. Most males are highly ornamented and fight for positions on a display ground, a *lek*, to attract females. Other males take a satellite role and a rare third type mimics female plumage and sneaks matings. Genomics revealed the super-gene controlling the complex pattern of morphs. This discovery would have been a fantasy even ten years ago.

Chapter 7
The wisdom of crowds

Self-organization

There is something mysterious about the way a flock of thousands of birds seems to manoeuvre as one, or the way a school of fish divides and reforms as it moves around a predator. Similarly the busy activity of ants leaving and returning to a nest catches our attention. While we can see the behaviour of individual animals as they move and interact, there is a level of organization and apparent coordination that is visible at the level of the flock, school, or nest. Recent work on collective animal behaviour shows that relatively simple behaviours by each animal acting individually can together produce complex, emergent behaviours larger than the parts. This self-organization occurs despite the limited cognitive abilities of individual animals, and despite each animal's limited access to global information, or the 'big picture'.

Such self-organization rules allow social insects and other animals to make complex choices such as how to allocate workers optimally to food sources, how to choose the best nest site from many possible sites, or how to build complex structures on scales far bigger than the individual animals. The phenomenon has been termed *swarm intelligence*. Simple rules and responses also appear to underlie collective motion of animal flocks and schools.

24. A murmuration of thousands of starlings.

Collective motion

Shortly before sunset, flocks of tens of thousands of European starlings (*Sturnus vulgaris*) fly in aerobatic displays called *murmurations* (Figure 24). The flocks swirl and morph, transforming from, say, a teardrop shape into a vortex, and then into a long rope, making spontaneous, synchronized turns as if of one mind. An early 20th-century British ethologist, Edmund Selous, was mystified as to how such big flocks could be so beautifully coordinated.

The explanation came from mathematical models and computer simulations. The collective motion of flocks of birds (and other animal groups) can be modelled as a flock of *self-propelled agents*. The coordinated formations do not require 'leaders'. Each agent (animal) bases its own movement decisions on the positions, orientations, motion (or change in motion) of a few near neighbours. Using this local information, each agent follows simple rules to keep close but not too close, and align with and match the velocity of neighbours. While each agent reacts only with its neighbours, these interact in turn with their own neighbours so a change in movement by an agent on one edge of the flock can ripple across it. The resulting simulated flocks behave convincingly like real ones, changing shape, dividing and reforming.

The theories have been tested by observations of murmurations of starlings flying over Rome. An interdisciplinary project has used stereoscopic photography and digital image processing to reconstruct the positions, tracks, and velocities of individual birds. The results have supported the basic theoretical idea—that local interactions between birds can explain the flock movement—but the real shape of the flocks is more like a sheet rather than the bulbous shapes we see them as from the ground. Another unexpected finding is that the flocks are densest towards the edges, and that flocks can differ more than two-fold in their density. This has lead researchers to propose that the birds respond to the nearest six or seven neighbours, rather than only neighbours within a certain radius as models had often assumed.

The speed and orientation of the starlings are highly correlated across the flock, no matter how big the flock, which accounts for the flock's synchronized movements. When birds in one part of the flock are disturbed (by seeing a predator such as a hawk, for example), their individual response in change of direction and/or speed transmits as a wave across the flock, amplifying the change of direction by just a few individuals initially. Transmission of information in this way creates an effective sensory range for the flock far greater than the perception of any individual.

Predation may also be one of the selection pressures leading to schooling in fish. Fish schools divide and reform around a predator, always just out of reach. The geometry may be driven in part by the *selfish herd*, each fish avoiding being the one nearest the predator (incidentally, this is another idea proposed by W.D. Hamilton). Fish use their lateral line organs, as well as vision, for detecting the position and movement of neighbours.

Some recurrent shapes are found in the collective motion of many different species of birds, fish, and other animals. Modelling can generate these by tweaking the strength of different interactions (Figure 25). In addition, only a few individuals with information about the direction of food or migration route are needed to influence the whole flock or school.

Collective decisions

Social insects give us perhaps the most extreme examples of the power of self-organization: although hundreds of thousands or even millions of individuals are involved in a colony, activity is not commanded from the centre. Instead, complex colony behaviours are built from the bottom up by multiple interactions between individual colony members, each with a miniature brain, using simple rules to respond to local information. In this way, a group of workers can collectively tackle tasks far beyond the abilities of any one individual. The resulting patterns, such as an intricate nest at a scale thousands of times larger than any individual, emerge from these multiple non-linear interactions between individuals. Many of the interactions involve pheromones.
A colony of social insects is a genetically related unit, so there is likely to be colony-level selection on the efficiency and optimality of the responses and rules used by the individuals of a colony.

Ant trails

Ant trails result from positive feedback. When a scout ant worker finds food (for example, cake at a picnic), the ant returns to the nest, laying down a *pheromone trail* from a scent gland. Other ants from the colony follow the pheromone trail out to the food source, laying down their own pheromone on the way back,

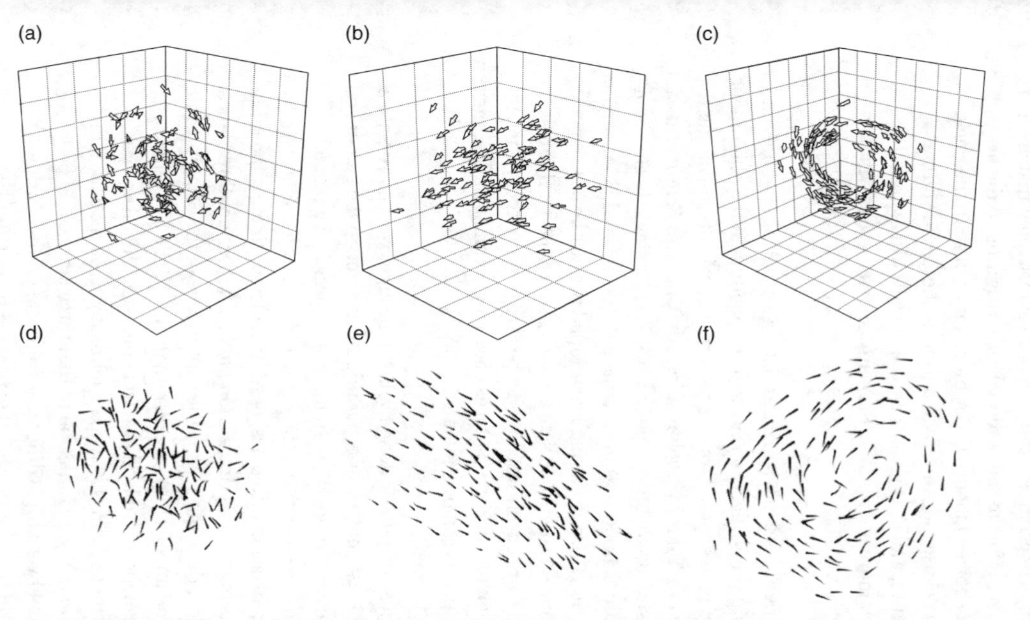

25. Modelling collective animal movements. (a) With only attraction and repulsion, disordered swarms like those of mosquitoes are created. (b) Adding some alignment gives a 'torus' of individuals rotating around a hollow core. (c) As alignment increases, cohesive 'polarized' groups start to form, like bird flocks and fish schools. Below, photographs of fish shoals show these transitions (d, e, f).

reinforcing the trail. When the food runs out, the ants no longer reinforce the trail and it fades away as the pheromone evaporates. This simple mechanism can lead to a sophisticated matching of ant forager effort to the potential resources around the nest. Given a choice of food sources, more ants will go to the richer source because more ants will lay more pheromone on the return trail, and ants given a choice of trails will choose the one with more pheromone. Similarly, shorter routes will be reinforced more.

Computer models of virtual ants laying virtual 'pheromone' and following simple decision rules for responding to the pheromone trail can recreate the foraging patterns of real ants, including the huge swarm raids by a colony of South American *Eciton* army ants comprising up to half a million workers. In a single twelve-hour raid more than 30,000 prey items may be captured. Some *Eciton* species hunt dispersed single arthropod prey while other species hunt 'concentrated' food sources such as wasp nests. If the dispersed or concentrated food types are put into the model, the different characteristic patterns of raiding columns of the different *Eciton* species can be simulated.

Two-way traffic flows along trails can be high, with more than a hundred fast-moving ants per minute, leading to the risk of collisions and congestion. In many ant species, the rate of collisions is reduced and the capacity of the trail is increased by self-organization into lanes, optimizing traffic flow. In the raiding columns of the army ant, *Eciton burchelli*, there are three lanes: inbound laden ants occupy the central lane of the column while outbound ants form a lane on either side. While a two-lane arrangement would lead to fewer collisions, the three-lane arrangement seems likely to be more stable. In models, Iain Couzin and Nigel Franks found that three lanes could be generated by a simple difference in turning rates (during interactions with other ants) between inbound and outbound ants.

Like ants, walking humans also form lanes spontaneously where there is two-way traffic such as on a pedestrian pavement. Pheromone trails are not involved, however, and nor do models of the type used for fish and bird groups explain our behaviour. Instead, it seems that humans' self-organization into lanes is based on unconscious heuristic cognitive processes—similar to the way we catch a ball—without consciously thinking how we do it. These unconscious processes adapt our walking speeds and directions in response to visual information about the distance of obstructions, including other people, ahead of us. Multiple lanes result as individual people actively seek a free path through the crowd. At high crowd densities, the models show that smooth flows break down into crowd turbulence, characteristic of situations leading to trampling incidents in crowd disasters.

Termite nests

Termites are master builders, constructing elaborate mud structures up to 30 metres in diameter. These structures are thousands of times larger than the individual workers, which are just millimetres long. Inside the 5-metre high nest of *Macrotermes* termites is the intricate network of air ventilation ducts which air-condition the mound, as well as the specialized chambers for the fungus gardens and for the brood, and a royal chamber for the king and queen (Figure 26).

However, there is no blueprint or top-down management of construction. Instead, these complex, species-specific shapes and structures emerge from simple rules and local interactions by individual termite workers. For example, the blind termite workers appear to build the royal chamber of *Macrotermes subhyalinus*, responding to a volatile pheromone diffusing from the queen. Workers deposit their mud load at a certain threshold concentration away from the queen, building the chamber wall and roof at that distance, following her contours. When the queen gets bigger, the distance marking the threshold will move out

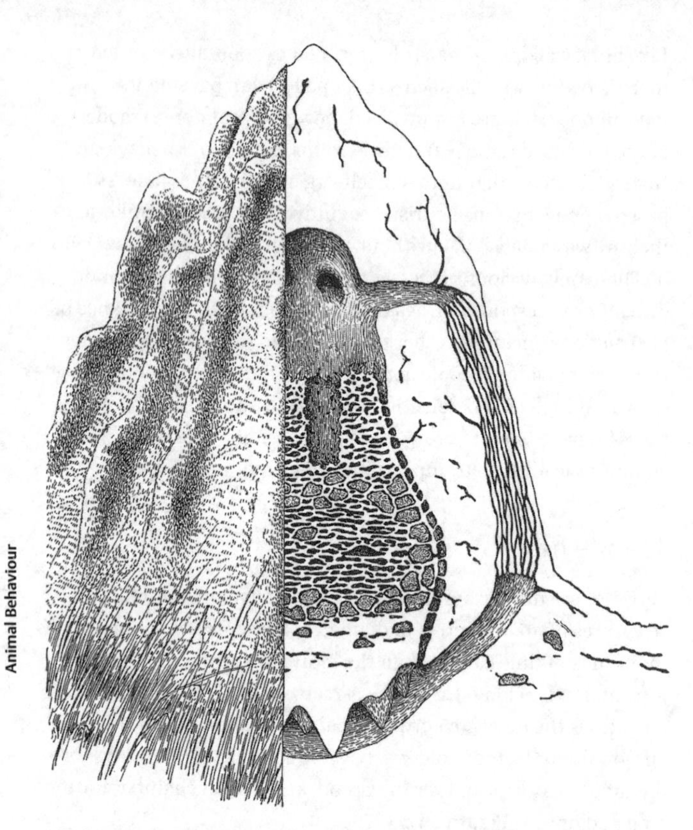

26. A *Macrotermes* termite mound. A vertical section (right) reveals a complex network of ventilation and living chambers for different functions. The 5-metre tall mud structure is constructed by blind workers each only 0.005 metres long.

further and the termite workers will remodel the walls to make a bigger chamber.

The termite workers also respond to local configurations of mud and pheromones laid down by other workers. For example, a pile of mud impregnated with a *cement pheromone* may stimulate the next termite to deposit more mud, triggering in turn another building action by that termite or any other worker in the colony. The self-organized construction of the complex elements of the termite mound—for example, a pattern of pillars and walls—can be explained by mathematical models. The models incorporate threshold responses of termites to cement and trail pheromones, diffusion and fading of pheromones. Other factors include responses to existing structures and other cues such as the air currents the structures create.

Honeybee democracy

In the late spring, a strong colony of honeybees (*Apis mellifera*) divides by swarming, with the current queen flying off with about half the workers, leaving the remainder in the old nest with a daughter queen. The swarm of several thousand bees, with the queen hidden at its centre, gathers like a beard on a tree branch nearby. The swarm delegates just a hundred or so of the older bees to fly off and scout for candidate new nest cavities in hollow trees. A scout bee looks for a hole and flies in, judging the cavity's size by walking about inside. An excellent site will offer the right volume for the colony to store enough honey to last a winter, as well as protection from predators by being high from the ground with a small, defensible south-facing opening.

When a scout returns, she does a waggle dance (Chapter 5) on the surface of the swarm. The angle of her dance gives the direction and the duration of the waggle run gives the distance to the site of the cavity. The perceived quality of the site is reflected by the strength of her dance shown by the number of times she repeats the waggle dance. For an excellent site, she will do more than a hundred encores. Other scouts return having found other sites. Each scout attempts to recruit other bees, with more being recruited by stronger waggle dances. The recruited bees are uncommitted until they too have inspected the site themselves and return to the swarm to dance.

Tom Seeley and colleagues discovered that given a choice of sites, the bees usually succeed in choosing the best site as the new nest cavity. How do the swarming honeybees decide? They use disagreement and competition, not compromise. The groups of scouts for each site compete to become the first to reach a quorum (which seems to be about 150 scout bees currently visiting a site). Excellent sites reach the quorum threshold faster as their scouts have stronger waggle dances and continue to visit and dance more often, creating a positive feedback loop. Negative feedback for poor sites comes from a constant dance decay rate which leads to less persistent dancing for poor sites. Nest scouts also send inhibitory 'stop' vibration signals by head-butting waggle-dancing scouts reporting other sites, causing them to stop dancing sooner. The combination of positive and inhibitory effects means that sites with more advocates will recruit even more rapidly, with a snowballing effect on the number of scouts recruited.

When the quorum has been reached, which can take up to three days, 'piping' scouts run over the swarm to activate the bees for lift off. Bees 'pipe' by buzzing their wings and pressing their abdomen onto another bee to transmit the vibration signal. Suddenly, the swarm, with the queen at its centre, takes off in the direction of the new nest. Scouts already at the nest cavity release a chemical signal (a pheromone) to guide the final landing of the swarm.

The process appears to lead to good decisions. An important feature is that each scout assesses the site before she dances. Seeley likens the process to a town meeting at which all scouts are well informed, can 'debate', and can vote.

Natural selection and collective behaviour

The behaviour of individual animals and the rules of interaction, from which the collective behaviours emerge, have evolved under natural selection just like other behaviours. For example, natural selection will have selected for particular responses to pheromones on a trail which lead to the observed behaviours. In social insects, natural selection on collective behaviour acts at the level of the colony. Colonies with workers behaving more optimally than other colonies will produce more reproductives (queens) for the next generation.

It is easy to remember that each bird in a flock has sophisticated cognitive abilities. However, in social insect self-organized systems too, each of the 'simple units'—such as a bee responding to local conditions—has its own brain with millions of neurons and numerous sensory inputs, despite its small size. Each social insect individual is a sensitive and versatile processing unit, with responses affected by its genes, experience, and context. A worker bee will respond to at least twenty different signals and integrate tens of different kinds of cues. Her behavioural state will be influenced by her age and experience as well as the behaviour of her fellow nestmates (Chapter 3).

Applications of self-organization models

Swarm intelligence, with many brains applied to a problem, is also the way that humans develop open source software. Swarm intelligence may have applications in business and organizations for better decision-making by groups. Studies of human groups suggest there may be advantages in ensuring the groups are composed of individuals with diverse opinions—another reason for encouraging diversity among employees. Other key features promoting better decision-making by groups include encouraging independence of opinion, truthful reporting, and mechanisms to avoid hierarchies and domination of discussion.

Modellers quickly realized that the principles of self-organization emerging from studies of social insects in particular could be applied in other situations. Ant-colony-optimization and control algorithms inspired by social insects' cooperative foraging have been used to solve routing problems for data in communication networks such as telecoms and the internet. Like the colonies that inspired them, networks based on these algorithms show flexibility and robustness in dynamic environments. The ways that animals self-organize to form flocks or schools have been used to inform the programming of robots for autonomous robot swarms.

The insights from the modelling of human crowd behaviour are being used to help us design public buildings, such as cinemas, that can be more quickly and safely evacuated in an emergency.

Chapter 8
Applying behaviour

In this together

Our behaviour as humans has profound effects on the lives of other animals. This chapter explores some of the ways that we can use an understanding of animal behaviour to reduce our conflict with animals as we compete with them for food and space in the global environment. We can also attempt to make life better for our domesticated animals. Every way of studying animal behaviour—from neuroscience, development, learning, to behavioural ecology—can play a part.

Reducing conflicts between humans and animals

One of our biggest conflicts with animals is caused by competition for food. This conflict includes direct competition with animals we consider pests, which feed on growing or harvested crops, as well as other animals that are impacted inadvertently. A better understanding of how animals behave can offer less damaging ways to avoid these conflicts.

The potential of using insect pheromones to control insect pests by manipulating their behaviour was recognized many decades before the first pheromone could be identified (Chapter 2). Adult moths do no harm: the caterpillars do the damage—as the 'worm' in the apple. One successful control tactic is to release synthetic female sex pheromone so that the confused males cannot find the real females, resulting in fewer fertilized eggs and fewer caterpillars. This mating disruption is used worldwide, for example to protect apples in the USA, tomatoes in Mexico, and aubergines (eggplants) in Pakistan. The technique has minimal environmental impact. This is because, in contrast to pesticides, pheromones are species-specific and only small, non-toxic quantities are used for mating disruption, so only the pest moths are affected. Spiders and other natural predators and parasitoids are left unharmed, leaving the ecosystem intact.

Long term, pheromones might offer ways of controlling the mice and rats that cause big food losses once grains have been harvested. We know that pheromones are very important in the social interactions and reproduction of these nocturnal animals. Current control largely uses poisons. However, rodents are hard to poison because of learned taste aversion: if a poison causes illness, the animal later avoids the bait. In addition, the use of poisons gives rise to animal welfare and environmental concerns.

Not all pests are small. African elephants raid villagers' crops in Kenya, prompting novel experiments to repel them. Biologists noticed that elephants avoided acacia trees containing beehives. African honeybees are aggressive and the elephants have learnt to run away from the sound of angry bees. The elephants have an alarm call, a deep 'rumble', which alerts family members to the threat of bees. The farmers and biologists experimented with a 'beehive fence' around the crops. Hollow beehives were suspended from wires between posts and linked in such a way that an elephant trying to sneak between poles would disturb beehives on either side, causing the bees to come out and sting (Figure 27). Elephants learned quickly and avoided crops protected with the beehive fences. An additional advantage was that villagers could also harvest honey and wax from the beehives.

27. Kenyan farmers setting up a beehive fence to protect crops from elephants. The thatch will shade the box-like beehive.

In North America and Europe, honeybees, famous for their navigation skills, have started losing their way back to their hives, apparently because their memory has been affected. The cause may be pesticides such as *neonicotinoids* used to control pest insects. Although bees are not the target of these pesticides, the chemicals get into the pollen and nectar that the bees collect. While these low levels of pesticide do not kill the bees, they seem to affect the bees' behaviour. Together with other factors (including diseases and other types of pesticides), neonicotinoids may be contributing to a decline in the pollinator populations of honeybees, as well as declines in other bees, such as bumblebees.

Just as honeybees are victims of collateral damage from pesticides, several species—most notably, dolphins—are the unintended victims of fishing with *gillnets*. Dolphins become trapped when they try to take fish from these nets. Mortality in gillnets is the most pressing threat to the conservation of many dolphins and other small cetaceans. Because dolphins communicate with whistle sounds and use sonar for echolocation to find prey and navigate, many fisheries with dolphin problems have taken advantage of this sensitivity to sound, experimenting with 'pingers', battery-powered sound emitters attached to the nets. Pingers emit a sound that dolphins avoid. Gillnet fisheries off the Pacific and Atlantic coasts of the USA have shown 50–60 per cent drops in unintended catches of dolphins and porpoises since pingers were made mandatory. This drop has been maintained over a decade, with no sign that the pingers are losing their effectiveness due to habituation.

Behaviour and animal conservation

A better understanding of animal behaviour—including mating systems, imprinting, migration, and interactions with other species—can be an important part of attempts to conserve endangered animals.

Kakapos (*Strigops habroptilus*) are an endangered species of flightless parrot in New Zealand. Captive breeding programmes had problems because sex ratios were out of balance—too many males were being produced. An understanding of concepts from behavioural ecology revealed the solution. Rich food was being fed to female kakapos to encourage them to lay. The food triggered sex allocation mechanisms which had long evolved to produce more males in times of food plenty, as bigger males would gain more mates in the kakapo's highly competitive mating system. With control of supplementary feeding, normal sex ratios returned.

The behavioural sexual imprinting on humans, described by Konrad Lorenz (Figure 9), has been a problem with captive breeding programmes for endangered birds. The solutions have been ingenious. For example, at whooping crane projects in North America, human helpers are disguised in head-to-toe bird-suits and use a crane's-head hand-puppet when feeding the young. Later, the young birds are taught the 1,200-mile migration route behind a microlight aeroplane. In many species, captive-reared animals have to be taught to fear predators, including humans.

The endangered giant panda (*Ailuropoda melanoleuca*) has been notoriously difficult to breed in captivity. Intensive study of the behaviour of pandas in the wild and at the breeding centre in Sichuan, China, has led to improved techniques yielding greater success in breeding pandas for release. For example, the pandas are frequently swapped between pens to expose them to each other's scent and allow them to exchange important messages related to individual identification, sex and reproductive status, reproductive maturity, and competitive status. The scientists report that this appears to increase sexual motivation and reduce aggression prior to the mating introduction.

In 1979, the large blue butterfly (*Maculinea arion*) became extinct in the UK, even in nature reserves. The cause was a mystery until scientists understood the complicated lifecycle of the butterfly. Patient studies showed that the caterpillars were picked up by a particular ant species (*Myrmica sabuleti*) and taken back into their nest. Chemical cues persuaded the ants that the caterpillars were lost ant larvae. Inside the nest, the butterfly caterpillars mimicked the smell and sounds of queen ants, and rewarded their hosts by eating ant larvae. The cause of the butterfly's extinction turned out to be the loss of sheep grazing the land: the grass grew too long for the particular ant species to thrive, so the butterfly's lifecycle was broken. The solution was to bring back the sheep, wait for the ants to return, and re-introduce the butterfly: now it is thriving.

Global human-induced changes

By far the greatest threats to animal conservation from human activity are massive natural habitat loss and fragmentation, largely due to agriculture, and anthropogenic global climate change. These effects are on such large scales that local conservation interventions are completely inadequate. Instead, behavioural scientists can make a significant contribution by studying the likely effects of these habitat and climate changes on animals' behaviour and survival, and campaigning on the issues.

Environments have always changed; the difference now is the fast rate and extent of change. For some species, a flexible behavioural change may be the first response to a changing environment. For example, over recent decades of warming, the great tits in woods near Oxford have adjusted their egg laying times to track the earlier emergence of oak leaves and the caterpillars that form a key part of the food for the nestlings. However, great tits in the Netherlands have not adjusted egg laying dates for the warmer springs, and with the increasing mismatch to caterpillar arrival, their populations are declining.

For other species, for example those with limited thermal tolerances or those dependent directly or indirectly on other species, there may be limits to how they can adapt. A hummingbird dependent on a mountain flower for nectar can move up the mountain year by year as the plants grow at higher, cooler altitudes over generations of rising temperatures, but once the flower and bird reach the top of the mountain, there's nowhere to go.

In the longer term, genetic changes in behaviour may be selected for to increase survival in a changing environment. For example desert cactus-living fruitflies may respond to climate change by selection on the timing of activity driven by their circadian clock genes, becoming active at cooler times of day. For many species, change may not be achieved in time to avoid extinction.

Along with climate change, rising anthropogenic carbon dioxide levels in the atmosphere cause another hazard: about a third of the carbon dioxide dissolves in the oceans making them more acidic. This change will affect the behaviour and physiology of marine organisms that have existed for fifty million years or so at the same stable acidity. When tested at the acidity levels based on current projections of conditions in 2100, the sense of smell of marine organisms is disrupted, leaving worms and crustaceans no longer responding to sex pheromones, fish failing to recognize the smell of predators, and marine larvae failing to find the right habitat. (It is important to note, though, that these experiments did not gradually change the pH over a few generations.) If as bad as feared, the effects on plants and animals will have repercussions throughout the food webs that produce the fish and crustaceans we harvest from the sea, quite apart from the impacts on marine life generally.

Welfare in captive animals

How can an understanding of animal behaviour inform the way we treat animals? Our interest in animal welfare rests on the widely held belief that non-human animals have feelings and experience emotions in ways that plants or non-animate objects do not. However, we do not know how animals feel or whether they are conscious. This led Tinbergen to suggest that though animals might have feelings, these feelings could not be studied because they could not be measured. However, we are reminded by Marian Dawkins, one of the pioneers of a scientific approach to animal welfare, that this problem is not unique to animals. We have similar problems with human consciousness, as we cannot know directly what someone else is feeling. Instead, each of us has to assume other humans have feelings and experiences like our own on the basis that they are like us. We assume that what other people tell us they are feeling is a reasonable indirect measure of what they are experiencing. We can also use non-verbal cues given by facial expressions and behaviours. Similarly, with animals, we can use their behaviour and other cues to search for indicators of what they are feeling.

Consciousness remains one of the toughest unsolved problems in biology. How the billions of neurons in our brains create human consciousness is still a mystery. However, it is worth separating two kinds of consciousness. First, there is *access consciousness*, the ability to think and reason, studied by cognitive ethologists interested in the intellectual abilities of animals and how these compare with those of humans (Chapter 4). Second, there is *phenomenal consciousness*, sometimes called *sentience*, which is the immediate sensation of pain or pleasure. Animal welfare science has focused on phenomenal consciousness, asking if animals can feel pain and experience negative emotions, such as fear or hunger, which we call suffering; and positive emotions, defined as experiences animals wish to repeat, which we call pleasure. Even using the word 'emotion' can be problematic, as it tends to imply much more than a physical sensation when experienced by humans.

Other mammals and birds do appear to sense pain. Only recently have the specific sense neurons for sensing damaging stimuli, *nociceptors*, been found in fish. Fish behavioural responses to substances that are painful to humans suggest that they may feel pain. It is possible that some invertebrates do too. The complex behaviours and learning abilities of octopus mean that they and other cephalopods are protected in European Union legislation on animal experiments as if they were vertebrates (Figure 28). However, while questions remain as to whether any other animal consciously experiences pain as we do, that uncertainty should not lessen our desire to reduce potential pain to other animals.

Leaving aside questions of animal consciousness, improving animal welfare can be achieved by focusing on maintaining animal health and giving animals what they want. Research programmes based on these measures can help legislators make evidence-based decisions on how best to improve animal welfare. A pragmatic approach on these lines would help all animals, as

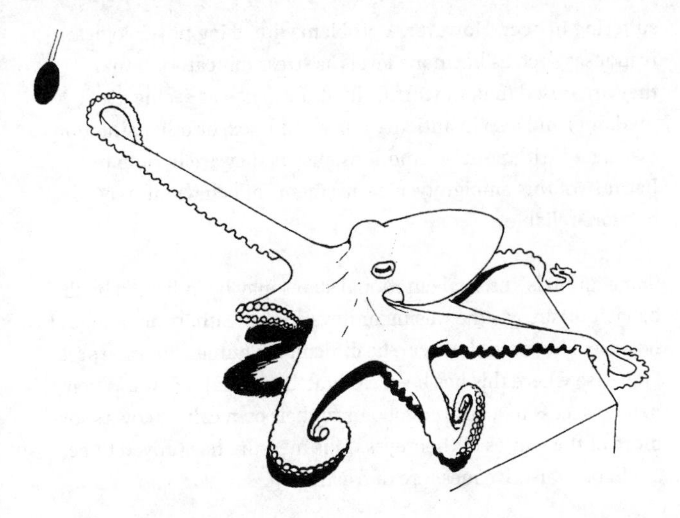

28. An octopus showing a cautious approach to a stimulus it has learnt to associate with a negative experience.

animals do not need to be 'clever' or have sophisticated intellectual abilities to experience pain or hunger.

Based on human emotions, animal welfare scientists have devised many ways of 'asking' animals what they want. Human emotions have three separate components. The first is our physiological response, including a rise in heart rate and release of hormones such as adrenaline. Second, we may give outward behavioural signs such as grimaces or sounds. Third, humans experience emotions such as fear or anger. Animal welfare science explores how valid it is to use the first two components to deduce what the animal might be experiencing. Using the word 'emotion' does not mean that animals necessarily feel things in the same ways as we do.

Raised levels of the 'stress' hormone *cortisol* in deer chased by dogs helped support the suggestion that hunting with dogs causes suffering in deer. However, a problem with using physiological responses such as hormone levels as stress indicators is that they are raised in both fearful situations (such as seeing a predator) and also in anticipation of food, sex, or other situations associated with positive emotions, just as they are in humans. Because of this ambiguity, measurements of behaviour may be more reliable.

Some animals' internal emotional states may be indicated by their behaviour such as the raising or lowering of feathers or changes in posture. However, these can be difficult for humans to interpret. One case where this has been calibrated is in dairy cows. When denied access to a desired food or to their own calves, cows show more of the whites of their eyes. This measure has proved to be a reliable objective measure of frustration.

Rather than trying to imagine what a chicken 'wants', we can ask chickens what they prefer, by offering choices in a behavioural experiment. The initial rise of scientific animal welfare in the 1960s was in response to concerns about conditions in factory farming, prompted in large part by Ruth Harrison's 1964 book *Animal Machines*. A United Kingdom government committee reasoned that chickens' feet would be more comfortable on battery cage floors made with thick wire. But when the chickens were 'asked' in an experiment, they chose to stand on floors made with thin wire.

Animals can also be asked how much they want something, how motivated they are. Rats will press a lever many more times to reach a cage with other rats than to access an empty cage, suggesting that social interactions are important to them. In the wild, mink swim to hunt prey in rivers. Caged mink will push doors, even when heavily weighted, to gain access to a swimming bath but not to access an empty space. Preventing access to the bath raised their urinary cortisol levels as much as preventing access to food. Animals can also be asked which experiences they dislike. Sheep quickly learnt to avoid a corridor that led to a room they associated with some shearing methods.

Asking the animals is important because, as with the battery chicken cages, our views of animal welfare and 'naturalness' may not reflect the animals' own choices. For example, 'free-range' chickens rarely leave their huts to range in open fields unless there are trees to forage under. This perhaps reflects their evolutionary past as jungle fowl which risked predation in the open. Cows may prefer to be indoors rather than out on pasture, especially when it rains.

Better animal welfare also leads to better outcomes for farmers. For example, calves taken away from their mother show abnormal behaviours, searching in vain for their mother's teats. These behaviours are reduced if the calves are fed frequently from an artificial teat, and they gain weight much more quickly than if fed from a bucket just twice a day. Part of the benefits may be due to the calf's sucking on the artificial teat, which stimulates release of digestive hormones that benefit the calf.

Improved welfare is also important for animals used in laboratory experiments. Mice kept in social groups in enriched environments with nesting material provide better models for studying human diseases (Figure 29). For example, mice housed in groups respond better to chemotherapy, perhaps in ways similar to the effects of social support on human responses to treatments.

However, including animal health as another indicator of welfare is important, as there may be a balance to be made when considering overall welfare. Dogs, like humans, want chocolate and other sweet foods and, like us, risk heart problems from being overweight. Free-range chickens may have poorer health than those indoors because they are more prone to parasite infections. Thus finding the right level of welfare sometimes involves trade-offs. Identifying the wants of domesticated pigs revealed

29. Giving nesting materials to laboratory mice helps them naturally regulate their body temperatures to a comfortable level, leading to less-stressed mice.

the benefits of providing a peat moss area for rooting for food, a separate dunging area, and an area where a sow about to give birth can collect leaves and straw to build a nest. These features reduce the frustrations and better meet the wants of the pigs, motivations that remain strong despite generations of domestication. However, providing these features needs to be balanced against some poorer health outcomes, especially neonatal survival, compared with conventional pens.

Understanding companion animals

For most of us, interactions with *companion animals* (pets) such as dogs and cats are our most common direct contact with animals. In evolutionary terms, our association is very recent, perhaps 15,000 years for dogs and 10,000 years for cats.

Molecular evidence suggests that dogs evolved from the wolf, *Canis lupis*, though where this occurred (and how many times) is still an active area of research. Initially, less-fearful wolves may have approached nomadic human encampments to scavenge. Gradually the association reached the stage of some pups being reared by people and the process of artificial selection for tameness and sociability could begin. Selection for tameness involves many genes affecting emotions, affiliative behaviour, and social communication. Attachment to humans, which develops in the young puppy, is illustrated by the way that dogs approach their owner when distressed or when they use the owner as a secure base, often looking back when exploring an unfamiliar space. From early puppyhood, dogs, unlike wolves, return our gaze. Like human babies, dogs can pick up our pointing or gaze shifts as cues to find objects. Dogs look at us to gain information about unfamiliar objects or events. A key element in the training of dogs is that they find human interaction and attention rewarding in itself. Simple learning allows dogs to learn combinations of tasks and situations so they can help us as seeing-eye or hearing dogs.

The ancestor of the domestic cat, the wildcat, *Felis silvestris*, is solitary and defends solitary territories, unlike the social ancestors of all other domesticated animals. A possible scenario is self-domestication: when agriculture started in the Fertile Crescent of the Middle East, some wildcats may have followed the mice that exploited the year-round food resources associated with humans. The ability to live around people gave advantages, and the divergence from their 'wild relatives' into domestication may have occurred by natural selection. When the domestic cat goes feral and lives in the wild, unlike other felids it lives in loose social groups of related females that raise their kittens together. This cat–cat social behaviour may have evolved before the closer interaction with humans as pets. Some of the signals directed towards humans are derived from kitten–mother communication (kneading, meowing), while others are adapted from signals used in cat social cohesion such as purring, cheek rubbing, and tail-up signalling.

Serious study of the behaviour of domestic dogs and cats, and their interactions with people, is relatively young. The recent focus on companion animal behaviour is welcome, given the importance of these animals in our lives.

What animals will there be to watch in future?

The animal behaviours introduced in this book and covered in countless engrossing television documentaries depend on their species' survival. Some animals, such as rats, mice, pigeons, and our companion animals, have adapted to sharing our lives and, as a result, they will prosper. We don't know which other animals will survive the changes made by humans, in particular the destruction of habitats and the impacts of global climate change.

Unless we want amazing animals to exist only in old documentaries, we need to act. The solutions involve behaviour, but the behaviour under scrutiny is our own behaviour as consumers and voters; and our influence on the behaviours of banks, businesses, and politicians worldwide. Change is possible.

References

Where possible I have chosen papers that are open access or freely available. Searching Google Scholar will often find the papers, sometimes on the scientist's own website. If you are not successful at first, try a Google search with the exact title in quotation marks. See also the books in 'Further reading': most of the classic studies will be mentioned there with full references.

Chapter 1: How animals behave (and why)

From hunting to ethology

Abramson, C.I.C. (2009). A study in inspiration: Charles Henry Turner (1867–1923) and the investigation of insect behavior. *Annual Review of Entomology*, 54: 343–59.

Galef, B.G. (1996). The making of a science. In L.D. Houck & L.C. Drickamer (eds), *Foundations of animal behavior*. Chicago: University of Chicago Press, pp. 5–12.

Meercats

Clutton-Brock, T. & Manser, M. (2016). Meerkats: cooperative breeding in the Kalahari. In W.D. Koenig & J.L. Dickinson (eds), *Cooperative breeding in vertebrates*. Cambridge: Cambridge University Press, pp. 294–317.

Primitively eusocial wasps

Gadagkar, R. (2011). Science as a hobby: how and why I came to study the social life of an Indian primitively eusocial wasp. *Current Science (Bangalore)* 100: 845–58.

Tinbergen's questions

Bateson, P. & Laland, K.N. (2013). Tinbergen's four questions: an appreciation and an update. *Trends in Ecology and Evolution*, 28: 712–18.

Crickets silenced

Pascoal, S. et al. (2014). Rapid convergent evolution in wild crickets. *Current Biology*, 24: 1369–74.

Fossil behaviour

Benton, M.J. (2010). Studying function and behavior in the fossil record. *PLoS Biology*, 8: e1000321.

Dance, A. (2016). Prehistoric animals, in living color. *Proceedings of the National Academy of Sciences USA*, 113: 8552–6.

Integrating the four questions

Schaal, B. & Al Aïn, S. (2014). Chemical signals 'selected for' newborns in mammals. *Animal Behaviour*, 97: 289–99.

Chapter 2: Sensing and responding

Bats and moths

ter Hofstede, H.M. & Ratcliffe, J.M. (2016). Evolutionary escalation: the bat–moth arms race. *Journal of Experimental Biology*, 219: 1589–602.

Sensory inputs

Catania, K.C. (2012). Evolution of brains and behavior for optimal foraging: a tale of two predators. *Proceedings of the National Academy of Sciences USA*, 109: 10701–8.

Sakurai, T., Namiki, S., & Kanzaki, R. (2014). Molecular and neural mechanisms of sex pheromone reception and processing in the silkmoth *Bombyx mori*. *Frontiers in Physiology*, 5: 125.

Neural circuits

Brembs, B. (2015). Watching a paradigm shift in neuroscience. *The Winnower*, 3: e142737.171063.

Domenici, P. et al. (2008). Cockroaches keep predators guessing by using preferred escape trajectories. *Current Biology*, 18: 1792–6.

Skals, N. et al. (2005). Her odours make him deaf: crossmodal modulation of olfaction and hearing in a male moth. *Journal of Experimental Biology*, 208: 595–601.

Solis, M.M. & Perkel, D.J. (2005). Neuroethology. *Encyclopedia of the life sciences*. Chichester: Wiley.

Drosophila circuits

Manoli, D.S., Meissner, G.W., & Baker, B.S. (2006). Blueprints for behaviour: genetic specification of neural circuitry for innate behaviors. *Trends in Neurosciences*, 29: 444–51.

Hormones and behaviour

Bridges, R.S. (2015). Neuroendocrine regulation of maternal behavior. *Frontiers in Neuroendocrinology*, 36: 178–96.

Maruska, K.P. & Fernald, R.D. (2013). Social regulation of male reproductive plasticity in an African cichlid fish. *Integrative and Comparative Biology*, 53: 938–50.

Stevenson, P.A. & Rillich, J. (2012). The decision to fight or flee—insights into underlying mechanism in crickets. *Frontiers in Neuroscience*, 6: 118.

Parasites

Hughes, D.P. (2014). On the origins of parasite-extended phenotypes. *Integrative and Comparative Biology*, 54: 210–17.

Libersat, F. & Gal, R. (2014). Wasp voodoo rituals, venom-cocktails, and the zombification of cockroach hosts. *Integrative and Comparative Biology*, 54: 129–42.

Chapter 3: How behaviour develops

Nature and nurture

Mameli, M. & Bateson, P. (2011). An evaluation of the concept of innateness. *Philosophical Transactions of the Royal Society B*, 366: 436–43.

Marler, P. (2004). Innateness and the instinct to learn. *Anais da Academia Brasileira de Ciencias*, 76: 189–200.

Imprinting

Bateson, P. (2011). Imprinting. *Scholarpedia*, 6: 6838.

Lévy, F. & Keller, M. (2008). Neurobiology of maternal behavior in sheep. *Advances in the Study of Behavior*, 38: 399–437.

Slagsvold, T. et al. (2002). Mate choice and imprinting in birds studied by cross-fostering in the wild. *Proceedings of the Royal Society of London Series B*, 269: 1449–55.

Genes and behaviour

Sokolowski, M.B. (2010). Social interactions in 'simple' model systems. *Neuron*, 65: 780–94.

Evolution, co-opted genes, and gene regulation

Carroll, S.B., Prud'Homme, B., & Gompel, N. (2008). Regulating evolution. *Scientific American*, 298 (May): 60–7.

Duncan, E.J., Gluckman, P.D., & Dearden, P.K. (2014). Epigenetics, plasticity, and evolution: how do we link epigenetic change to phenotype? *Journal of Experimental Zoology B*, 322: 208–20.

Robinson, G.E., Fernald, R.D., & Clayton, D.F. (2008). Genes and social behavior. *Science*, 322: 896–900.

Social influences

Ben-Shahar, Y. et al. (2002). Influence of gene action across different time scales on behavior. *Science*, 296: 741–4.

Maruska, K.P. & Fernald, R.D. (2013). Social regulation of male reproductive plasticity in an African cichlid fish. *Integrative and Comparative Biology*, 53: 938–50.

Phenotypic plasticity

Avise, J.C. & Mank, J.E. (2009). Evolutionary perspectives on hermaphroditism in fishes. *Sexual Development*, 3: 152–63.

Bachtrog, D. et al. (2014). Sex determination: why so many ways of doing it? *PLoS Biology*, 12: e1001899.

Simpson, S.J., Sword, G.A., & Lo, N. (2011). Polyphenism in insects. *Current Biology*, 21: R738–49.

Bird song as a model system

Bolhuis, J.J., Okanoya, K., & Scharff, C. (2010). Twitter evolution: converging mechanisms in birdsong and human speech. *Nature Reviews Neuroscience*, 11: 747–59.

Nottebohm, F (2005). The neural basis of birdsong. *PLoS Biology*, 3: e164.

Effects between generations

Freund, J. et al. (2013). Emergence of individuality in genetically identical mice. *Science*, 340: 756–9.

Sachser, N., Kaiser, S., & Hennessy, M.B. (2013). Behavioural profiles are shaped by social experience: when, how and why. *Philosophical Transactions of the Royal Society of London B*, 368: 20120344.

Play

Bateson, P. (2014). Play, playfulness, creativity and innovation. *Animal Behavior and Cognition*, 2: 99.

Sharpe, L.L. (2005). Play fighting does not affect subsequent fighting success in wild meerkats. *Animal Behaviour*, 69: 1023–9.

Chapter 4: Learning and animal culture

All animals learn

Menzel, R. (2013). Learning, memory, and cognition: animal perspectives. In G. Galizia & P.-M. Lledo (eds), *Neurosciences—from molecule to behavior: a university textbook*. Berlin: Springer.

Learning gives flexibility

Hollis, K.L. & Guillette, L.M. (2015). What associative learning in insects tells us about the evolution of learned and fixed behavior. *International Journal of Comparative Psychology*, 28: 1–18.

Hiding food for later

Brown, M.F. & Cook, R.G. (eds) (2006). *Animal spatial cognition: comparative, neural, and computational approaches*. Available online at: <www.pigeon.psy.tufts.edu/asc/>.

Woollett, K. & Maguire, E.A. (2011). Acquiring 'The Knowledge' of London's layout drives structural brain changes. *Current Biology*, 21: 2109–14.

Learning bees

Giurfa, M. (2013). Cognition with few neurons: higher-order learning in insects. *Trends in Neurosciences*, 36: 285–94.

Menzel, R. (2012). The honeybee as a model for understanding the basis of cognition. *Nature Reviews Neuroscience*, 13: 758–68.

Zhang, S., Si, A., & Pahl, M. (2012). Visually guided decision making in foraging honeybees. *Frontiers in Neuroscience*, 6: 88.

Social learning

Aplin, L.M. et al. (2014). Experimentally induced innovations lead to persistent culture via conformity in wild birds. *Nature*, 518: 538–41.

Galef, B.G. & Laland, K.N. (2005). Social learning in animals: empirical studies and theoretical models. *Bioscience*, 55: 489–99.

Laland, K.N. (2008). Animal cultures. *Current Biology*, 18: R366–R370.

Teaching

Thornton, A. & McAuliffe, K. (2006). Teaching in wild meerkats. *Science*, 313: 227–9.

Cumulative culture

Yuhas, D. (2012). *Sharing the wealth (of knowledge): cumulative cultural development may be exclusively human* [Online]. New York: Scientific American. <www.scientificamerican.com/article/sharing-the-wealth-of-kno/> (accessed 25 February 2016).

Tool use

Biro, D., Haslam, M., & Rutz, C. (2013). Tool use as adaptation. *Philosophical Transactions of the Royal Society B*, 368: 20120408.

Kacelnik, A. (2009). Tools for thought or thoughts for tools? *Proceedings of the National Academy of Sciences USA*, 106: 10071–2.

Seed, A. & Byrne, R. (2010). Animal tool-use. *Current Biology*, 20: R1032–9.

What are they thinking?

Emery, N.J. & Clayton, N.S. (2001). Effects of experience and social context on prospective caching strategies by scrub jays. *Nature*, 414: 443–6.

Penn, D.C. & Povinelli, D.J. (2007). On the lack of evidence that nonhuman primates possess anything remotely resembling a 'theory of mind'. *Philosophical Transactions of the Royal Society B*, 362: 731–44.

Pepperberg, I.M. (2006). Cognitive and communicative abilities of Grey parrots. *Applied Animal Behaviour Science*, 100: 77–86.

Shettleworth, S.J. (2010). Clever animals and killjoy explanations in comparative psychology. *Trends in Cognitive Sciences*, 14: 477–81.

Chapter 5: Signals for survival

Getting the message across

Wyatt, T.D. (2015). How animals communicate via pheromones. *American Scientist*, 103: 114–21.

Honest signals

Számadó, S. (2011). The cost of honesty and the fallacy of the handicap principle. *Animal Behaviour*, 81: 3–10.

Honeybee dances

Couvillon, M. (2012). The dance legacy of Karl von Frisch. *Insectes Sociaux*, 59: 297–306.

Tautz, J. et al. (2004). Honeybee odometry: performance in varying natural terrain. *PLoS Biology*, 2: e211.

Vervet monkey alarm calls

Ducheminsky, N., Henzi, S.P., & Barrett, L. (2014). Responses of vervet monkeys in large troops to terrestrial and aerial predator alarm calls. *Behavioral Ecology*, 25: 1474–84.

Yong, E. (2015). How reliable are psychology studies? *The Atlantic*. August. <www.theatlantic.com/science/archive/2015/08/psychology-studies-reliability-reproducability-nosek/402466/>.

Interspecies cooperation communication

Spottiswoode, C.N., Begg, K.S., & Begg, C.M. (2016). Reciprocal signaling in honeyguide-human mutualism. *Science*, 353: 387–9.

Vail, A.L., Manica, A., & Bshary, R. (2013). Referential gestures in fish collaborative hunting. *Nature Communications*, 4: 1765.

Exploitation of signals

Haynes, K.F. et al. (2002). Aggressive mimicry of moth pheromones by a bolas spider: how does this specialist predator attract more than one species of prey? *Chemoecology*, 12: 99–105.

Lewis, S.M. & Cratsley, C.K. (2008). Flash signal evolution, mate choice, and predation in fireflies. *Annual Review of Entomology*, 53: 293–321.

Ryan, M.J. (2011). Replication in field biology: the case of the frog-eating bat. *Science*, 334: 1229–30.

Chapter 6: Winning strategies

Winning strategies

Birkhead, T. & Monaghan, P. (2010). Ingenious ideas: the history of behavioral ecology. In D.F. Westneat & C.W. Fox (eds), *Evolutionary behavioural ecology*. New York, Oxford University Press, pp. 3–15.

Parental care and mating systems

Emlen, S.T., Wrege, P.H., & Smith, L. (2004). Size dimorphism, intrasexual competition, and sexual selection in Wattled Jacana

(*Jacana jacana*), a sex-role-reversed shorebird in Panama. *The Auk*, 121: 391–403.

Sperm competition

Birkhead, T. (2010). How stupid not to have thought of that: post-copulatory sexual selection. *Journal of Zoology*, 281: 78–93.

Explaining altruism

Bourke, A.F. (2014). Hamilton's rule and the causes of social evolution. *Philosophical Transactions of the Royal Society B*, 369: 20130362.

Emlen, S. & Wrege, P. (1989). A test of alternate hypotheses for helping behavior in white-fronted bee-eaters of Kenya. *Behavioral Ecology and Sociobiology*, 25: 303–19.

Mumme, R.L. (1992). Do helpers increase reproductive success? *Behavioral Ecology and Sociobiology*, 31: 319–28.

Russell, A. et al. (2007). Helpers increase the reproductive potential of offspring in cooperative meerkats. *Proceedings of the Royal Society of London B*, 274: 513–20.

Evolution of eusociality

Boomsma, J.J. (2009). Lifetime monogamy and the evolution of eusociality. *Philosophical Transactions of the Royal Society B*, 364: 3191–207.

Evolutionary arms races

Davies, N.B. (2011). Cuckoo adaptations: trickery and tuning. *Journal of Zoology*, 284: 1–14.

The future of behavioural ecology

Lamichhaney, S. et al. (2016). Structural genomic changes underlie alternative reproductive strategies in the ruff (*Philomachus pugnax*). *Nature Genetics*, 48: 84–8.

Owens, I.P.F. (2006). Where is behavioural ecology going? *Trends in Ecology and Evolution*, 21: 356–61.

Zuk, M. & Balenger, S.L. (2014). Behavioral ecology and genomics: new directions, or just a more detailed map? *Behavioral Ecology*, 25: 1277–82.

Chapter 7: The wisdom of crowds

Self-organization (and other sections)

Couzin, I.D. (2009). Collective cognition in animal groups. *Trends in Cognitive Science*, 13: 36–43.

Krause, J., Ruxton, G.D., & Krause, S. (2010). Swarm intelligence in animals and humans. *Trends in Ecology and Evolution*, 25: 28–34.
Theraulaz, G. et al. (2003). The formation of spatial patterns in social insects: from simple behaviours to complex structures. *Philosophical Transactions of the Royal Society B*, 361: 1263–82.

Collective motion

Cavagna, A. et al. (2010). Scale-free correlations in starling flocks. *Proceedings of the National Academy of Sciences USA*, 107: 11865–70.

Collective decisions

Boomsma, J.J. & Franks, N.R. (2006). Social insects: from selfish genes to self organisation and beyond. *Trends in Ecology and Evolution*, 21: 303–8.

Ant trails

Fourcassié, V., Dussutour, A., & Deneubourg, J.-L. (2010). Ant traffic rules. *Journal of Experimental Biology*, 213: 2357–63.
Moussaïd, M., Helbing, D., & Theraulaz, G. (2011). How simple rules determine pedestrian behavior and crowd disasters. *Proceedings of the National Academy of Sciences USA*, 108: 6884–8.

Honeybee democracy

Seeley, T.D., Visscher, P.K., & Passino, K.M. (2006). Group decision making in honey bee swarms: when 10,000 bees go house hunting, how do they cooperatively choose their new nesting site? *American Scientist*, 94: 220–9.

Chapter 8: Applying behaviour

Pheromones

Wyatt, T.D. (2015). How animals communicate via pheromones. *American Scientist*, 103: 114–21.

Elephants and bees

King, L.E., Douglas-Hamilton, I., & Vollrath, F. (2011). Beehive fences as effective deterrents for crop-raiding elephants: field trials in northern Kenya. *African Journal of Ecology*, 49: 431–9.

Neonicotinoids

Pisa, L.W. et al. (2014). Effects of neonicotinoids and fipronil on non-target invertebrates. *Environmental Science and Pollution Research*, 22: 68–102.

Dolphins and 'pingers'

Dawson, S.M. et al. (2013). To ping or not to ping: the use of active acoustic devices in mitigating interactions between small cetaceans and gillnet fisheries. *Endangered Species Research*, 19: 201–21.

Conservation and behaviour

Buchholz, R. (2007). Behavioural biology: an effective and relevant conservation tool. *Trends in Ecology and Evolution*, 22: 401–7.

Caro, T. (2007). Behavior and conservation: a bridge too far? *Trends in Ecology and Evolution*, 22: 394–400.

Thomas, J.A., Simcox, D.J., & Bourn, N.A.D. (2011). Large blue butterfly. In P.S. Sourae (ed.), *Global Re-introduction Perspectives 2011*. Gland: IUCN. <https://portals.iucn.org/library/efiles/documents/2011-073.pdf>.

Zhang, G., Swaisgood, R.R., & Zhang, H. (2004). Evaluation of behavioral factors influencing reproductive success and failure in captive giant pandas. *Zoo Biology*, 23: 15–31.

Climate change and behaviour

Leduc, A.O. et al. (2013). Effects of acidification on olfactory-mediated behaviour in freshwater and marine ecosystems: a synthesis. *Philosophical Transactions of the Royal Society B*, 368: 20120447.

Wong, B.B.M. & Candolin, U. (2014). Behavioral responses to changing environments. *Behavioral Ecology*, 26: 665–73.

Animal welfare

Dawkins, M.S. (2006). A user's guide to animal welfare science. *Trends in Ecology and Evolution*, 21: 77–82.

Dawkins, M.S. (2015). Animal welfare and the paradox of animal consciousness. *Advances in the Study of Behavior*, 47: 5–38.

Fraser, D. (2008). Understanding animal welfare. *Acta Veterinaria Scandinavica*, 50: S1.

Companion animals

Bradshaw, J.W.S. (2015). Sociality in cats: a comparative review. *Journal of Veterinary Behavior: Clinical Applications and Research*, 11: 113–24.

Driscoll, C.A., Macdonald, D.W., & O'Brien, S.J. (2009). From wild animals to domestic pets, an evolutionary view of domestication. *Proceedings of the National Academy of Sciences USA*, 106: Suppl 1, 9971–8.

Miklósi, Á. & Topal, J. (2013). What does it take to become 'best friends'? Evolutionary changes in canine social competence. *Trends in Cognitive Science*, 17: 287–94.

Further reading

Citizen science

As well as suggestions for further reading, I would like to mention some of the many ways to get actively involved through citizen science and contribute to our understanding of animal behaviour. Zooniverse (<www.zooniverse.org>) offers one of the biggest collections of projects, such as identifying animals for WildCam Gorongosa in images from Mozambique or helping identify bat calls in Bat Detective. There are many projects involving birds in your own garden such as the Cornell Bird Lab's Feederwatch (<http://feederwatch.org/>) in the USA. If you are interested in bees, in the UK there is Beewatch (<http://bumblebeeconservation.org/get-involved/surveys/beewatch/>) and in North America there is Bumble Bee Watch (<www.bumblebeewatch.org>). There are likely to be similar projects of various kinds which match your own interests and location (see <www.scistarter.com>).

Magazines, videos, podcasts, and blogs

Animal behaviour is featured in many magazines in print and online, such as *BBC Wildlife*, *Discover*, *New Scientist*, and *Scientific American*.

There are links to many videos on behaviour on the Animals Behave blog <www.AnimalsBehave.org>.

The BBC has many web resources focused on animal behaviour: <www.bbc.co.uk/nature/wildlife> or <www.bbc.com/earth/uk>.

To hear scientists speaking about their work in animal behaviour, listen to engaging podcasts in the BBC radio programmes on natural history: <www.bbc.co.uk/programmes/p02rrm3j>. These include series like *In Our Time*, for example, on behavioural ecology: <www.bbc.co.uk/programmes/b04tljk0>; and *The Life Scientific*, for example an interview with Nicky Clayton who works on bird cognition: <www.bbc.co.uk/programmes/b017cd0v>; or Nick Davies on cuckoo behaviour: <www.bbc.co.uk/programmes/b07gfft2>. From the USA there is a wide range of interesting podcasts and videos on NPR's *Science Friday*: <www.sciencefriday.com>, for example on firefly flash communication: <www.sciencefriday.com/?s=fireflies>.

You will find many good blogs related to animal behaviour on the web. Among many others, I enjoy the blogs and tweets of @edyong209; @GrrlScientist; @Mammals_Suck; @JoanStrassmann; and @carlzimmer. I blog at <www.AnimalsBehave.org> and tweet behaviour links myself from @BehavingAnimals.

Books

Four textbooks on animal behaviour cover the field from slightly different approaches. All are recommended:

Alcock, J. (2013). *Animal behavior: an evolutionary approach*, 10th edn. Sunderland, MA, Sinuaer.

Bolhuis, J.J. & Giraldeau, L.A. (eds) (2005). *The behavior of animals: mechanisms, function, and evolution*. Oxford, Blackwell.

Davies, N.B., Krebs, J.R., & West, S.A. (2012). *An introduction to behavioural ecology*, 4th edn. Chichester, Wiley-Blackwell.

Manning, A. & Dawkins, M.S. (2012). *An introduction to animal behaviour*, 6th edn. Cambridge, Cambridge University Press.

A well-written textbook covering animal behaviour from the neuroscience angle is:

Galizia, G. & Lledo, P.M. (eds) (2013). *Neurosciences—from molecule to behavior: a university textbook*. Berlin, Springer.

Richard Dawkins' explanations of genes, evolution, and behaviour have been highly influential. Marek Kohn gives an illuminating account of these evolutionary ideas and some the people involved, including Richard Dawkins, R.A. Fisher, J.B.S. Haldane, W.D. Hamilton, and others.

Dawkins, R. (2016). *The selfish gene: 40th anniversary edition.* Oxford, Oxford University Press.
Kohn, M. (2004). *A reason for everything: natural selection and the British imagination*. London, Faber and Faber.

Marlene Zuk brings a lively discussion of concepts in animal behaviour much wider than the book's title would suggest:

Zuk, M. (2011). *Sex on six legs: lessons on life, love, and language from the insect world*. New York, Houghton Mifflin Harcourt.

Like many of the leading scientists studying animal behaviour, Tim Birkhead came via a fascination with birds. He introduces bird behaviour in these books:

Birkhead, T. (2012). *Bird sense: what it's like to be a bird*. London, Bloomsbury.
Birkhead, T., Wimpenny, J., & Montgomerie, B. (2014). *Ten thousand birds: ornithology since Darwin*. Princeton, Princeton University Press.

Chapter 1: How animals behave (and why)

Burkhardt, R.W. (2005). *Patterns of behaviour: Konrad Lorenz, Niko Tinbergen, and the founding of ethology*. Chicago, University of Chicago Press.
Clutton-Brock, T. (2008). *Meerkat manor: flower of the Kalahari*. London, Phoenix (Orion).
Kruuk, H. (2003). *Niko's nature: the life of Niko Tinbergen, and his science of animal behaviour*. Oxford, Oxford University Press.
Lorenz, K.Z. (1964). *King Solomon's ring*. London, Methuen.
Tinbergen, N. (1951). *The study of instinct*. Oxford, Clarendon Press.
Tinbergen, N. (1974). *Curious naturalists*, revised edn. Harmondsworth, UK, Penguin.
von Frisch, K. (2014). *Bees: their vision, chemical senses, and language*. Ithaca, Cornell University Press.

Chapter 2: Sensing and responding

Berkowitz, A. (2016). *Governing behaviour: how nerve cell dictatorship and democracies control everything we do*. Cambridge, MA: Harvard University Press.

Hughes, D.P., Brodeur, J., & Thomas, F. (eds) (2012). *Host manipulation by parasites*, Oxford: Oxford University Press.
Simmons, P. & Young, D. (2010). *Nerve cells and animal behaviour*, 3rd edn. Cambridge, Cambridge University Press.
Stevens, M. (2013). *Sensory ecology, behaviour, and evolution*. Oxford, Oxford University Press.

Chapter 3: How behaviour develops

Arney, K. (2016). *Herding Hemingway's cats: understanding how our genes work*. London, Bloomsbury.
Catchpole, C.K. & Slater, P.J.B. (2008). *Bird song: biological themes and variations*, 2nd edn. Cambridge, Cambridge University Press.
Dawkins, R. (1986). *The blind watchmaker*. London, Longman/Penguin.
Luck, M. (2014). *Hormones: a very short introduction*. Oxford, Oxford University Press.
Nelson, R.J. & Kriegsfeld, L.J. (2017). *An introduction to behavioral endocrinology*, 5th edn. Sunderland, MA, Sinuaer.
Slack, J. (2014). *Genes: a very short introduction*. Oxford, Oxford University Press.

Chapter 4: Learning and animal culture

Laland, K.N. & Galef, B.G. (eds) (2009). *The question of animal culture*. Cambridge, MA, Harvard University Press.
Pepperberg, I. (2009). *Alex & me: how a scientist and a parrot discovered a hidden world of animal intelligence—and formed a deep bond in the process*. New York, Harper.
Sanz, C.M., Call, J., & Boesch, C. (eds) (2013). *Tool use in animals: cognition and ecology*. Cambridge: Cambridge University Press.
Shettleworth, S.J. (2010). *Cognition, evolution, and behavior*, 2nd edn. Oxford, Oxford University Press.

Chapter 5: Signals for survival

Bradbury, J.W. & Vehrencamp, S.L. (2011). *Principles of animal communication*, 2nd edn. Sunderland, MA, Sinauer (and free web resources available <http://sites.sinauer.com/animalcommunication2e/>).

Cheney, D.L. & Seyfarth, R.M. (1990). *How monkeys see the world: inside the mind of another species.* Chicago, University of Chicago Press.

Maynard Smith, J. & Harper, D. (2003). *Animal signals.* Oxford, Oxford University Press.

Munz, T. (2016). *The dancing bees: Karl von Frisch and the discovery of the honeybee language.* Chicago, University of Chicago Press.

Stevens, M. (2016). *Cheats and deceits: how animals and plants exploit and mislead.* Oxford, Oxford University Press.

Tautz, J. (2008). *The buzz about bees: biology of a superorganism.* Berlin, Springer.

Wyatt, T.D. (2014). *Pheromones and animal behavior: chemical signals and signatures*, 2nd edn. Cambridge, Cambridge University Press.

Chapter 6: Winning strategies

Davies, N.B. (2015). *Cuckoo: cheating by nature.* London: Bloomsbury.

Davies, N.B. (2015). Cuckoos and their victims: an evolutionary arms race. Croonian Lecture, Royal Society. Video available at <https://royalsociety.org/events/2015/05/cuckoos-and-their-victims/>.

Davies, N.B., Krebs, J.R., & West, S.A. (2012). *An introduction to behavioural ecology*, 4th edn. Chicester, Wiley-Blackwell.

Chapter 7: The wisdom of crowds

Camazine, S. et al. (2001). *Self-organization in biological systems.* Princeton, Princeton University Press.

Hölldobler, B. & Wilson, E.O. (2009). *The superorganism: the beauty, elegance, and strangeness of insect societies.* New York, W.W. Norton.

Seeley, T.D. (2010). *Honeybee democracy.* Princeton, Princeton University Press.

Chapter 8: Applying behaviour

Berger-Tal, O. & Saltz, D. (eds) (2016). *Conservation behavior: applying behavioral ecology to wildlife conservation and management.* Cambridge, Cambridge University Press.

Bradshaw, J. (2012). *In defence of dogs*. London, Penguin.
Bradshaw, J. (2014). *Cat sense: the feline enigma revealed*. London, Penguin.
Broom, D.M. (2014). *Sentience and animal welfare*. Wallingford, CABI.
Dawkins, M.S. (2012). *Why animals matter: animal consciousness, animal welfare, and human well-being*. Oxford, Oxford University Press.
Miklósi, Á. (2014). *Dog behaviour, evolution, and cognition*, 2nd edn. Oxford, Oxford University Press.
Turner, D.C. & Bateson, P. (eds) (2013). *The domestic cat: the biology of its behaviour*, 3rd edn. Cambridge, Cambridge University Press.

Index

D

E

F

G

M

N

O

P

R

S

T

W

Z

GEOGRAPHY

A Very Short Introduction

John A. Matthews & David T. Herbert

Modern Geography has come a long way from its historical roots in exploring foreign lands, and simply mapping and naming the regions of the world. Spanning both physical and human Geography, the discipline today is unique as a subject which can bridge the divide between the sciences and the humanities, and between the environment and our society. Using wide-ranging examples from global warming and oil, to urbanization and ethnicity, this *Very Short Introduction* paints a broad picture of the current state of Geography, its subject matter, concepts and methods, and its strengths and controversies. The book's conclusion is no less than a manifesto for Geography' future.

> 'Matthews and Herbert's book is written- as befits the VSI series- in an accessible prose style and is peppered with attractive and understandable images, graphs and tables.'
>
> **Geographical.**

www.oup.com/vsi